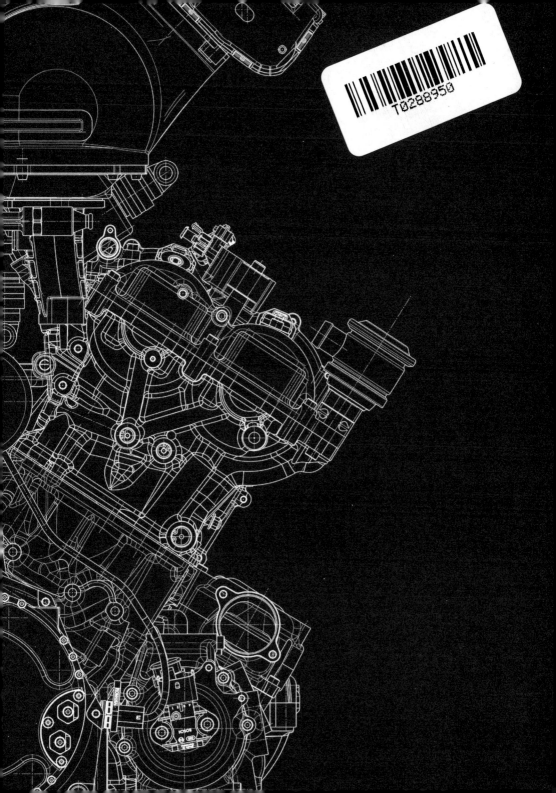

T0288950

THE AGE
OF COMBUSTION

THE AGE
OF COMBUSTION

NOTES ON AUTOMOBILE DESIGN

STEPHEN BAYLEY

CIRCA

First published in 2021 by Circa Press
©2021 Circa Press Limited and Stephen Bayley

Circa Press
50 Great Portland Street
London W1W 7ND
www.circa.press

ISBN 978-1-911422-13-6

All rights reserved. No part of this publication may be
reproduced or transmitted in any form or by any means,
electronic or mechanical, including photocopy, recording or
any other information storage and retrieval system, without
prior permission in writing from the publisher.

Printed and bound in Italy

Reproduction: Dexter Premedia
Design: Jean-Michel Dentand

CONTENTS

THE AGE OF COMBUSTION

Although I was brought up with automobile culture, it was a happy accident that I began to write about cars. I'd really been destined for less serious things.

The very first photograph of me shows a bouncy baby, not with a cuddly toy, but sitting on the vast reflector headlamp of my father's Georges Roesch Talbot. So stylishly improvident was my father, he perhaps overreached himself with the impressive Talbot, but I later discovered that natural equilibrium was restored because the Talbot, purchased used, had a piston missing. This had a retarding effect both on the car and, for a time, on my father's showmanship.

As an older child, it was more normal for me to be comfortably installed in the back of an interesting car than in a well-furnished drawing room. I saw the world through the Triplex fenestration of a Jaguar, a Humber or an Austin Westminster. Long before reading about Henry Ford's determination to escape the suffocating boredom of life on the farm by means of his gasoline buggy, I had intuited that cars offered an alternative reality to suburban ennui. Or perhaps I mean the sole reality. Soon we'll be away from here, step on the gas and wipe that tear away. Paul McCartney, a suburban boy, never wrote anything more moving.

Architecture became my subject, but then I went to a lecture by Nikolaus Pevsner, who explained that, so far as he was concerned, cars and architecture were designed to identical disciplines, the only distinction was that one was a 'mobile controlled environment' while the other was a 'static controlled environment'. Thus, the greatest architectural historian of them all confirmed my belief that it would not be intellectual slumming to take cars seriously.

Besides, I enjoyed the mischief involved with my one-time academic colleagues, all outraged by my wild assertions that a '57 Chevy might reasonably be considered on the same terms as a '57 Dick Smith. To paraphrase Tom Wolfe, if Brancusi is art, so are cars. With no disrespect to either Brancusi or Smith, artists I enjoy, I still believe it. And to paraphrase Wolfe again, to hell with Moholy-Nagy, whoever he is. Anyway, isn't Max Friz more interesting?

Raising cars to the level of art, even of applied art, grew into a preoccupation. In 1982, I became the first person to put a car in London's Victoria & Albert Museum. This was a 1947 Saab 92. Dark green. Cars had hitherto been ignored by the world's leading applied art museum because the dusty old antiquarian curators could not decide if they were metalwork or sculpture. Of course, they are both. Except, that is, if they are made of glass-reinforced plastic, like a Trabant.

Yet design as a whole was my real subject and I confess that I am not a keen driver nor an amateur of workshop activities. A slight stigma, a reputational stain, still attaches to those who profess an interest in motoring. But how can you study design without studying cars, since they are the most designed product of them all? For good or for bad, and I think mostly for good, cars demand more creative investment than any other product we buy. If you want to be intellectually responsible, it would be intellectually irresponsible not to attempt to understand the forces that create them and the powers we use to interpret them.

What follows here is an edited and revised selection of the monthly 'Aesthete' column I have written in *Octane* magazine for more than ten years. One way or another, it is perhaps the most consistent commentary on car design ever. And *Octane* is perhaps the world's outstanding classic car magazine whose global popularity confirms, if confirmation were needed, that classic cars perform extraordinary feats in our collective imaginations. Vehicles of desire? Maybe. Certainly, they represent a paradise that is lost. As Proust, something of a car man himself, put it *'Les vrais paradis sont les paradis qu'on a perdus'.*

Proust's countryman, the critic Sainte-Beuve, defined classic as something which is 'universal and permanent'. It's good to remember that as chilly, bloodless electricity takes over from the hot, smelly noisy Age of Combustion.

BRITISHNESS

Queen Elizabeth, Fidei Defender, 1954
Queen Elizabeth II in a Land Rover, Tobruk, 1954. When
Maurice Wilks, a member of the family that owned Rover,
and a collaborator on the development of the Whittle jet
engine, acquired an Army surplus American Jeep, he
decided he could improve it. Wilks' idea was presented
to the world at the 1948 Amsterdam Motor Show. It is a
near-perfect geometrical composition, using only
straight lines and very simple radii. There is no frivolous
decoration, nor anything not determined by purpose, but
so far from being banal and utilitarian, the Land Rover is
evocative and delightful.

The Prince of Wales in his birthday present, 1971
As Defender of the Faith, the future King Charles III also
needs to defend the British motor industry. This Aston-
Martin DB6 Volante was given to him by his mother as a
twenty-first birthday present, in 1969. However, the DB6
was a ham-fisted development of the elegant DB4, based
on an original design by Touring of Milan. The Kammback
tail in racing style is incongruous and a desperate nod
towards technological fashion. Even at its launch, in 1965,
commentators noted how dated the DB6 was. Prince
Charles loved it and had the petrol engine converted to
use bioethanol made from British grapes.

A Greek engineer in Birmingham, 1965
The Mini was the greatest, perhaps the only, triumph of
the forlorn British Motor Corporation. So, it is delightful
to note that its creator, Alec Issigonis, was a lugubrious
Greek from Smyrna, its body was influenced by the
Longbridge house-style established by an Argentine-
Italian, called Riccardo Burzi, and its nearest relation, in
engineering terms, was the curious German *Kleinwagen*
– bubble car – of the fifties. The Mini was technically
ingenious, commercially bold, ergonomically revolutionary,
fun to drive, and ineffably chic. It was the first small car to
transcend the stigma of class. And it became perhaps
the most admired and imitated design of all time.

INDUCTION

Jack Kerouac's *On the Road* (1957) has become a timeless classic, effortlessly transcending the druggy and solipsistic limitations of its Beat Generation milieu. It was a road trip, and his car was a '49 Hudson, and Kerouac wrote not on sheets, but on a continuous roll of paper, a device that surely aided his fluency. Here we read: Q. 'Where are we going?' A. 'I don't know, but I gotta go'.

Truly, we are at the end of an era. You are reading this in the last days of the Age of Combustion, an art historical period as precise, as meaningful and as productive of fascination and beauty as the Rococo or Baroque. Of torment and distress too. Perhaps every era has its scary Inquisitions.

Why 'Induction' and not an 'Introduction'? Because the four-stroke-cycle of the fuel-burning internal combustion engine was defined by Nikolaus Otto as induction, compression, ignition, exhaust. Induction is the beginning. That seems no less than a gloss on life itself. The more so in the interpretation of Keith Duckworth, designer of the most successful Grand Prix engine: 'suck, squeeze, bāng, blow'. For nearly one hundred and fifty years, the car has been, among the many other roles it adopted or accepted, a rich source of metaphor as it has sucked, squeezed, banged and blown its way around the planet.

Jack Kerouac's road trip is one of literature's finest acknowledgements of the car in culture. That quote above perfectly captures the automobile's absurd promise and its shocking betrayal: drivers are initially tempted by a suggestion of freedom, but are ultimately incarcerated in one way or another, whether physically or psychologically.

But driving a car is not just about travel. Perhaps it never really was. It is also the experience of friction and the discipline of gears, an awareness of penetrating the atmosphere, the thrill of speed, an opportunity to test your perceptions, nerves and reflexes. Of finding a vista of escape. Occasionally of confronting fear. And all of this excitement on a daily basis. A daily commute offers raw sensual experience which, a century ago, was the stuff of deranged fantasy to the Futurists.

The car also offers, as designers and consumers have discovered, in one of art history's best-ever double-acts, generous opportunities for public displays of taste, affluence and belief. Cars are, as Roland Barthes so very correctly said in his paean to the '57 Citroën, 'our cathedrals'.

That may be an understatement. But acts of worship are certainly involved in each. Barthes had seen the Citroën at the Paris Salon of 1955, one of the car industry's highly ritualised religious performances which may have put the cathedral image in his mind. But the car was called a DS which, pronounced the French way, sounds like 'goddess'.

In driving, you get a sense of controlling a machine as if a divine authority. In visiting a cathedral, the interaction is of an uncontaminated spiritual or intellectual kind. Long before 'cybernetics' was a coinage of Norbert Wiener's, in 1948, any car driver had known how humans integrate with mechanisms. In a bleakly digital age, we can now see that driving a car is a richly analogue experience ... in fact, the ultimate analogue experience, if you discount shooting yourself in the head with a Mossberg MC2c compact semi-automatic handgun with double-stack magazines and new ergonomic features.

What happens when you drive? A petrol-air mixture is inducted into cylinders through a quaintly complicated device called a carburettor and is then compressed by a reciprocating piston. John Ruskin disliked pistons because they reminded him of the intromittent binary sex which so disturbed him and it is true that, in a way, the Otto Cycle in many respects faithfully apes the Laws of Nature, with its mixture of vitality and futility.

And these Laws are by no means always rational. Professor Otto's piston reaches its maximum speed just an instant before it is violently decelerated to a standstill in what is poetically known as Top Dead Centre. There's another metaphor waiting to be exploited.

Professor Otto's Cycle is not, to be honest, an especially efficient one, even if thermodynamics has its own strange and compelling poetry and no one ever wants poets to be worried by efficiency. There is so much shameful waste heat in a combustion engine, radiators are required to transfer thermal energy from the metal engine to a liquid coolant. No purpose other than the survival of the cylinder-block is served by this.

Petrol does not burn, but its flammable vapours do. In fact, they explode. And this explosion, which creates the power stroke that ultimately allows you to break the speed limit, releases an excess of carbon dioxide, nitrogen oxides and various unburned hydrocarbons. A gallon of petrol weighs 7.3lbs and releases about 5.5lbs of carbon. Your District Nurse's apparently innocent Morris Minor was, for all its charm, a heavily disguised mobile environmental calamity. The filthy exhaust pipe might be compared to the anus.

But Professor Otto also offered a huge creative brief to designers, an artistic challenge to marry his machinery to meaningful forms. All that heat being dealt with by the radiator required an air intake which, in the radiator grille, became one of the car designer's favourite motifs: an orifice to be decorated and loaded with meaning. An orifice which replicates the mouth as surely as headlights replicate the eyes. Yes, of course, cars have faces.

And while no layman will ever be able to fathom what actually goes on in an AMD Ryzen Threadripper 3960X CPU, this month's most powerful computer chips, gearboxes and brakes are much easier to construe. Or, at least, easier to admire. This is, partly, because they talk to you: bytes are silent, cogs and callipers are not.

A gearbox is a power transmission system whose purpose is to adjust the speed of Professor Otto's mad heat engine to the speed of the vehicle. Its gears are little discs with teeth cut into them. Engine and gearbox are connected via a clutch, a mechanism allowing engagement and disengagement. The gearbox is operated by a lever; the clutch, most often, by a foot pedal.

Then you will need brakes to rein in all this dynamic chaos should you wish progress to come to a complete halt or merely for speed to be diminished. On her trial run in the first feasible car in Mannheim, in 1886, Bertha Benz had brakes which were not much more complicated than a wooden lever bound with leather which acted on the wheels to convert kinetic energy to thermal energy through the agency of friction, bringing Bertha to a juddering standstill. Modern brake pads are more complicated and may contain bronze, chalk, graphite, vermiculite, phenolic resin, steel, rubber, sand and aramid fibres. In this way was progress measured in the last century.

This bizarre mechanical circus, this theatre of physics, this catalogue of heat, noise, speed and waste proved deeply stimulating to art. While electric motors differ only in size, there are many formats of combustion engine, and each has its own character to stimulate the designer of the car it occupies.

While it is not inevitable that a little 850cc iron-block, pushrod four-stroke should inspire the style of a Morris Minor, it would, nonetheless, have been absurd if the little 1948 Morris had looked like a grotesque 1967 Cadillac Eldorado Fleetwood or a phallomorphic 1961 Jaguar E-type. Similarly, an operatic Ferrari V12 with six down-draught Weber carburettors breathing through polished trumpets would not have seemed at home in a car as cute and tiny as a Fiat Cinquecento. What I am saying is that Professor Otto's noisy circus positively demanded interpretation by designers.

But to understand the car it's also necessary to have a sense of the complex, not so say 'muddled', psychological realities which gave us the modern automobile. Not least, the company founded by Bertha Benz's husband was named after a woman: 'Mercedes' Benz was a tribute to the daughter of the marque's distributor in Nice. Tom Wolfe acknowledged this complexity when he said that cars are 'freedom, style, sex, power, motion, colour … everything'. But they are more than that.

Ford's Model T was the first practical car. It made the ordinary American a 'man enthroned', according to EB White in his beautiful 1936 essay 'Farewell, My Lovely', mourning the Model T's demise. Twenty years later, Hertz was offering enthronement for $7.85 a day to anyone who hired a car. The Hertz ads said, 'Rent it here, leave it there', the ultimate proposition of an American culture based on the mobility and ease which, for a wonderful moment, the automobile offered.

Yet, despite the revolution in global habits caused by the automobile – they changed the shape of cities according to Frank Lloyd Wright – the motor industry remains inherently conservative. The general arrangement of the car with its four wheels, seats in between and motor at one end or another, has not changed much since Mannheim.

In contrast to this conservatism, the romance of the car became intense. From *Huckleberry Finn* to *Grand Theft Auto*, via Kerouac and Bob Dylan's *Highway 61*, America read like a road epic. Europe too, but our poetics were different. America's most popular car was Henry's democratic Ford. Europe's most popular car was the Volkswagen, designed by Ferdinand Porsche, a willing helpmeet of Hitler's.

For the purposes of this Induction, America takes the driver's seat. Consider F Scott Fitzgerald, the great poet of ruined glamour and wasted promise. In 1920, flush with the advance from *This Side of Paradise*, he fired up his 1918 Marmon, bundled his loopy wife into the passenger seat and drove from Connecticut to Alabama, so Zelda could rediscover the peaches and biscuits of her Southern youth. They were looking for a lost Golden Age, which was to become the subject of *The Great Gatsby*. (In the book, a yellow Rolls-Royce plays an important part.)

Fitzgerald turned this eight-day journey into a series of articles, which appeared in the US *Motor* magazine, in 1924, eventually published in book form, in 2011, as *The Cruise of the Rolling Junk*. The reality was one of bust axles, blow-outs, lost wheel, and misdirections, since Zelda could not read a map. Additionally, Dr Jones' *Guide Book for Autoists* misled them: a route described as a navigable highway turned out to be the rocky bed of a dried-out stream. Find what metaphor you like here.

Scott and Zelda never found their Golden Age, but Fitzgerald could not let the fantasy go. He described 'an ethereal picture of how we would roll southward along the glittering boulevards of many cities, then, by way of quiet lanes and fragrant hollows whose honeysuckle branches would ruffle our hair with white sweet fingers'. That's what a Marmon could do for you.

On return, Zelda icily wrote 'the joys of motoring are more-or-less fictional'. But because they are fiction does not make them less compelling. Indeed, isn't the enchantment of an imagined journey to an impossible destination what fundamentally attracted us, as well as F Scott Fitzgerald, to cars?

Cars may not (quite) be art, but during the last century replaced art as a source of public guidance and instruction in matters of aesthetics. They usurped art's role. People learn about form, symbolism, colour, chiaroscuro and details from looking at cars, not from visiting art galleries. While, to an impressive extent, twentieth-century artists abandoned the pursuit of beauty, car designers took up the chase. Often very successfully.

Naturally, French intellos were among the first to recognise this. Besides Roland Barthes with his ineffable essay on 'The New Citroën' in his 1957 collection of essays, published as *Mythologies*, there was Gilbert Simondon, author of 1958's *Du mode d'existence des objets techniques*.

Here, in a bravura passage, he explains the formal evolution of designs as if they are an evolutionary process: 'Artefacts evolve using themselves as the point of departure; they contain the conditions for their own development. The structure of the object moves to match the future conditions in which it will be employed'. Indeed, they were an evolutionary process … until the Age of Combustion came to an end.

Simondon had not anticipated this degringolade, but he saw life in machines, evidenced by his congratulations to Britain's National Coal Board on its painstaking restoration of a Newcomen engine. 'There is', he said, 'something eternal in a technical schema.' Indeed. Even as the private car ceases to be the paramount consumer product, its extraordinary imagery will survive.

Car design, along with pop music and the movies, was one of the unique and defining activities of the twentieth century. Significantly, each was a collective activity with more than one auteur.

Consider the car's role in the cinema. The title *Rebel Without a Cause*, the 1955 film that made James Dean famous, was taken from an academic psychological study about alienated and disturbed youth. In the film, his car was a six-passenger 1949 Mercury Club Coupe with a 255 cu in (4.2 litre) flathead V8 … a car that later became a favourite of Californian hot-rodders and customisers, the inspiration of Tom Wolfe's famous remark.

A month before *Rebel* was released, a twenty-four-year-old Dean was killed in the Californian desert when his Porsche 550 RSK collided with a Ford station wagon. This lent Dean's lead sled a certain sinister posthumous cultishness as a memorial to destroyed youth, while enhancing Porsche's reputation for danger.

Alfred Hitchcock understood how cars convey meaning. His wife, Alma, chose the pretty, metallic ice-blue Sunbeam Alpine Series III that Grace Kelly drove in *To Catch a Thief*, the masterpiece romantic comedy-thriller. In export markets, the Alpine was intended to fit between the primitive MG TD and the more sophisticated Jaguar XK120.

It was also the perfect fit for the beautiful, but chilly, Grace Kelly character. Her stylish drive along the Haute Corniche with a soundtrack of wince-making squeals of cross-ply Dunlop tyres and scrunching gravel, with an immaculately dressed Cary Grant as a composed, but very nervous, cat-burglar passenger, is one of cinema's great car sequences.

Indeed, the name 'Alpine' expressed some of the potent mix of exotic yearnings felt by the British during the long, grim moment of postwar rationing. So influential was the film in creating a popular concept of Continental glamour, the views of Monte Carlo from the Corniche near La Turbie later became a visual cliché in car advertising.

Like *To Catch a Thief*, Mike Nichols' *The Graduate*, of 1967, was a superlative film based on a mediocre novel. It is one of the greatest rite-of-passage-coming-of-age movies ever, combining motifs of forbidden sex, rebellion, redemption, and an Alfa Romeo plays a leading role as a means of escape, all set to unforgettable sing-along music by Simon and Garfunkel.

The Alfa is a 1966 Series I Duetto with the classic 1.6 litre twin-cam four. Its body was drawn by Franco Martinengo of Pininfarina, and this early version has the distinctive curvaceous *coda di seppia* (cuttlefish) tail. Alfa's well-known frailties become a dramatic moment in the film when the Duetto runs out of petrol in the Californian desert because of a faulty fuel gauge.

But the great visual moment in the film is the Dustin Hoffman character, Benjamin Braddock, rushing over San Francisco's Bay Bridge on the way to frustrate the young Ms Robinson's wedding. The Rosso Corsa Alfa is seen from above, travelling from the Embarcadero to Yerba Buena Island (and also going the wrong way through the Gaviota Tunnel on US Route 101).

The speeding car somehow conveys a perfect expression of wistful longing. Additionally, the elegant Alfa confers on Hoffman-Braddock the keenest sense of eroticised style and nervous urgency. To judge how successful was the Alfa's role, simply imagine how different the mood would have been if instead it had been the District Nurse's Morris Minor.

A recent survey suggested that in the whole history of cinema, the most familiar car is a New York Ford Crown Victoria yellow cab. Manufactured between 1992 and 2011, it has featured nearly five thousand times in the movies. But the most famous cab of them all is the Checker A-series driven by Travis Bickle, the

Robert de Niro character in Martin Scorsese's greatest film, *Taxi Driver*, of 1976. Bickle, a brooding, foul-mouthed angry Vietnam veteran, sits in command of his own world, a scratched and yellow Perspex partition separating him from passengers sitting on dirty blue bottom-polished vinyl. Bickle's car is a 1974 A-series with the small-block 350 cu in (5.7 litre) Chevrolet V8 grumbling through a three-speed auto box. During the production run from 1959 to 1982, the Checker A-series was all but unchanged. Eternity, as Simondon suggested, might lie in a mechanical schema.

But what does the greatest movie franchise of them all tell us about cars? If William Lyons had not been so epically tight-fisted, James Bond's car would have been an E-type. The producers tried to blag freebie Jaguars, but Lyons was disinclined to deal. So, 007 got a DB5. (But that was only in *Goldfinger*, the third Bond film. No one remembers now, but Bond's first movie car was a Sunbeam Alpine, successor to Grace Kelly's.)

It's an assumption of pop-culture analysis that cars in the cinema add meaning and mystique to heroes and villains. Of course, Bond's creator, Ian Fleming, had already brought impressive amounts of brand-related snobbery to his masterpieces. His own taste for cars was altogether different: he drove Thunderbirds because he enjoyed their power and gadgets. Latterly, a Studebaker Avanti with an incongruous Home-Counties-style badge-bar. Meanwhile, the real-life spy Kim Philby drove a grim Humber from the Ministry of Defence carpool.

So, these are the precious brand values Connery brought to Aston Martin: sadism, sexism, snobbery, hedonism, refined violence, great good taste, a sharp, but relaxed, sense of style, gentle wit and amazing muscles. Really, you could not have a more valuable set of associations for fast cars. And here is your answer to the story of all cars in cinema: they perform roles. And they take you places never expected. And that, of course, is one definition of art.

There is poignancy here, as there always is when eras end.

Yet even today car designers spend more time in the future than you might imagine, corrupting mondain perceptions of time and of space. By the time a conventional new car is launched, its designers have already designed the replacement of its replacement. Thus, with an eight-year product life cycle, designers today are already sixteen years ahead of us. Or put it this way: the brand-new combustion-engine car you buy in 2021 was conceived circa 2005. Its future is now a thing of the past.

The promised flying car has not yet taken off. One of the first was envisaged by no less than Henry Ford himself who said, in 1940: 'Mark my words: a combination airplane and motor car is coming. You may smile, but it will come'. We are still smiling. Yet the vision persists. In every annual news cycle, endearing eccentrics emerge to claim they have perfected a flying car, only to disappear again.

In 1909, EM Forster wrote his short story 'The Machine Stops'. It's a gloomy tale. People live underground, paying tribute to a remote and scary technical entity known as The Machine. They no longer travel because they use video conferencing and text. Sound familiar?

Those disenchanted with the system are stigmatised as 'unmechanical', just as social-media-refusers are today stigmatised as neo-Luddites. And then do you know what happens? The Machine Stops. When the hot server farms in the Arizona desert melt because there is no more diesel or gasoline to power the generators that power the air-con they greedily devour, our machine will stop too.

The economist, John Maynard Keynes, said by about now we would all be so rich, no one would have to work. He was wrong too. The Poet Laureate of Dystopia, JG Ballard, predicted by about now, private cars would not be allowed on public roads. Instead, 'enthusiasts' would use them under psychiatric supervision in enclosed 'motoring parks'. Ballard was more nearly right: the only time you can actually use even a fraction of the potential of your ludicrous supercar is on a private track day.

The great adventure of the car is over. Never again will a cinnamon-linen-suited Detroit executive, with highly polished brogues, instruct his studio: 'I want that line to have a duflunky, to come across, have a little hook in it, and then do a rashoom or zong', as GM's Harley Earl once did.

Since Barthes's bold equivalence of cars with cathedrals, our reverence for the automobile has been adjusted. But they remain as the Sorbonne savant put it 'the best messenger of a world superior to that of Nature'.

Thing is, that message has now changed. No longer majestically enthroned, drivers are more nearly humiliatingly entombed in congestion. The promise of freedom has been broken. Dreams of escape are not today based on wheeled transport. Distant horizons no longer exist, and a road trip is an ordeal not a pleasure. Petrol stinks and adventure has been replaced by sorrow.

Some say the world will end in fire/Some say in ice.
Robert Frost, 1923

DETAILS

T43, a mechanical ballet, 1931

When the Bugatti T43 appeared, in 1931, Modernist architects, designers and artists were preoccupied with the beauty of machinery. The 'Tubist' painter Fernand Léger had made a film, *Ballet Mécanique*, which featured whizzing components. Marcel Duchamp declared the contents of a plumber's merchant to be 'art'. US Customs had insisted a Brancusi sculpture was an industrial component. And Ettore Bugatti returned the compliment by first modelling his engines in wood, to check that the style was right.

Buckminster Fuller's Dymaxion car, 1933

The racing driver, Francis Turner, was killed in 1933 while testing Buckminster Fuller's cynically under-developed prototype Dymaxion car, a vehicle that its designer held to be a herald of a technology-rich future of his own interpretation. Despite his blowhard prophecies, Fuller had no grasp of the engineering or aerodynamic fundamentals of car design. The ludicrous three-wheeled Dymaxion would lift at speed, rendering steering and brakes useless, as Turner discovered. The contrast between euphoric potential and dire reality in Buckminster Fuller's technocratic universe is revealing.

Mustang instruments, 1965

The 1965 Ford Mustang, an exercise in consumer psychology as much as vehicle engineering, included the option of a 'Rally-Pac'. The presence of this extra was indicated by two additional instruments in pods around the steering column, partially concealing the speedometer … perhaps not something an authentic rally driver would wish to do. But if Ford capo Lee Iacocca cared very little for ergonomics, he understood his customers with genius insight. Every option available on the Mustang calibrated not just engine speed or water temperature, but mysterious human desire itself.

THE MECHANICAL GRAVEYARD

What can we hear in the mechanical graveyard? Certainly, the ugly shriek of nibblers and jigsaws tearing old sheet metal. Perhaps the diminuendo of a once lively hot engine growing cold. I will never forget that, when he was working on the McLaren F1, Gordon Murray told me he studied Nick Mason's Ferrari 250 GTO so that he could understand the precise musical ticking that a fine engine made as its metals cooled and contracted. I don't think an electric pole-switched reluctance motor is ever likely to inspire similarly poetic curiosity.

What's on my mind is the dead or dying art of the engine. Artists once recognised the beauty of machinery. In 1924, the Cubist painter Fernand Léger made a now famous film called *Ballet Mécanique*. Inspired by trench memories of a First World War 75mm machine gun, Léger spoke of 'the magic of light on white metal'. His film, a masterpiece of abstract cinema, treated engines as things of mysterious beauty. (Less mysterious was the optional extra soundtrack, where wooden broomsticks were fed into electric fans to make a harrowing noise.)

Nineteen-twenty-four was the year when Bugatti's Type 35 appeared. Ettore Bugatti came from a family of artists, but he did not see his greasy mechanical calling as culturally inferior. It is said he modelled his engines first in wood to check that the proportions were correct. Certainly, they were hand-finished to such fine tolerances that gaskets were not necessary between block and head. And often parts of Bugatti's engines were given a guilloche finish, that fine, repetitive, incised shimmering pattern you get from engine-turning. So far as Ettore Bugatti was concerned, intense pleasure was to be had from merely looking at such a power unit.

Artistically speaking, although some will find this *nostalgie de la boue*, I feel a keen affection for BMC's A-series engine. A modest, humble thing with little or no pretension, it nonetheless achieved a sort of nobility as an uncomplicated statement of mechanical fact. Its specification was simple, perhaps even crude, with three bearings, pushrods and a shameless embrace of cast iron, but look at one now and it's hard not to feel a pang of elegy at the sight of a lost world. I especially like the green-painted rocker box. And, given patience, if you looked long enough at, for example, the A-series' ancient SU carburettor, you could intuit how it worked from scrutiny of springs and levers alone. Art must, after all, be instructive.

At a more elevated level, a good measure of the E-type Jaguar's generous deposits of visual drama belonged to the engine itself. The designers knew this. On early models a T-shaped tool was clipped to the footwell. This you needed to unfasten the vast bonnet which, tipping heavily and creakily forward, performed a highly self-conscious theatrical reveal of the noble antiquity that was the old twin-cam straight six. It stood there as proud as a temple. If the BMC A-series was the equivalent of architecture's Primitive Hut, the Jaguar six-cylinder was the Parthenon: all sophistication, detail and refinement.

This art has all but disappeared. As if in prudish denial of the raw mechanical facts of life, most modern engine bays are as veiled and shrouded as a woman in purdah (with all the associations of shame and chastity that word carries). Today, an engine says nothing visually, as if the manufacturer were embarrassed by its moving parts and seeks a disguise. Those sinister black plastic shrouds, the polite, stainless-finish highlights, the total denial of burning oil and erotically intromittent pistons! Today's engine wears a hijab. And it is just as dishonest and dispiriting.

As in all things, Porsche is different. Porsche engines have rarely been things of emphatic beauty. Indeed, open the hatch on a 911 and it looks no more exciting than a washing-machine second-hand parts depot. But art can take many forms. I know this because I have driven a Porsche 991 and don't expect ever to drive a more fabulous car. One reason for this? The noise. With great art, the water-cooled flat-six Porsche reproduces the raucous valve clatter, strange frequencies and warbling resonances of the old air-cooled engine. Actually, this is more than reproduction. It is glorious exaggeration.

Porsche's petrol engines might have one foot in the grave, but while we await the burial service, they make a glorious, glorious noise. The other sound in the mechanical graveyard is of gas escaping at high velocity. It's very beautiful and aesthetes must enjoy it while they can.

CARBURETTORS

'Carburettor' is a word now almost extinct in everyday life. Thus, it joins amarulence (bitterness) and sinapistic (containing mustard) in a treasury of neglect. As soon as I started thinking this, it occurred to me that I scarcely know what a carburettor is. Or I should say *was*.

The carburettor predates the automobile and, according to the *Oxford English Dictionary*, was first used in 1866, exactly twenty years before Benz's *Motorwagen* patent. A carburettor was an apparatus for passing hydrogen, coal gas, or air over a liquid hydrocarbon so as to produce what the lexicologists delightfully described as 'illuminating power'.

The names associated with illuminating power read like a glorious synopsis of history. Besides Benz, there were De Cristoforis, Bernardi, Daimler, Maybach, Lanchester, Skinner's Union, Dellorto, Holley and, of course, Weber. And there is a wonderful poetic vocabulary too, now fading into memory. Downdraught, butterfly valves, float chamber, choke, throttle jets, flooding and backfire.

I especially remember flooding. Although at the time I was completely ignorant of Bernoulli's Principle, the one that says, in flowing air, speed and pressure work inversely which governs the design of carburettors, I spent many unhappy hours under the bonnet of my old MG trying to balance its erratic twin SUs. I had not a clue what I was doing but sensed nonetheless that I was enjoying a meaningful relationship with my machinery.

And I think this is why carburettors are interesting. It's certainly why I am nostalgic for them. They evoke an age when you could look at an engine and read its function. My MG's A-series four cylinder was like a child's diagram of the laws of physics. You could see that air came in one side, got mixed with fuel on its way into the cylinder-block where, by way of a controlled explosion, it produced that illuminating power whose residue escaped as smelly hot gas out of the other side. It was elemental, even primal, and very satisfying, aesthetically speaking.

Carburettors spoke a visual language all of their own. The pleasant, rounded domes of the classic SU seemed to suggest English correctness, even a slight primness, since they rather disguised some of its functions. Altogether more feral were the Amal side-draught carburettors, the ones used on JAP-engined Cooper Formula Three cars and on Triumph Bonneville bikes. These seemed almost erotically assertive: a mechanical proposition with a sexual suggestion. Visually, they say raw power. You can almost hear them hiss.

Or look at a vast four-barrel Holley carburettor, as gross as a piece of military equipment, the sort you might find on a US muscle car with a Muncie four-on-the-floor and lake pipes painted matt white. How could such gigantic apparatus be anything other than American? Over-specified and over-large, even separated from the engine it illuminates, a Holley says four hundred cubic inch gurgling, molecule-bashing V8.

But most evocative of all is a Ferrari V12 with its martial row of six pairs of glittering twin downdraught Webers. Their fabulously arrogant air trumpets strut themselves and suck in the atmosphere like the most rampant gigolo. Such a display (and it *is* a display) of seductive, magnificent and ridiculous excess could only be Italian. Actually, it could only be from Emilia-Romagna, just as an SU could only be from Birmingham.

In 1991, Ferrari organised an exhibition of exceptional splendour in Florence's Belvedere with its greatest cars in vast Plexiglas boxes on the hillside, floodlit against a Renaissance backdrop. It was quite the thing, but what I enjoyed the most was the exhibition catalogue, which had gravure illustrations of the greatest Ferrari engines. Whenever I am quizzed about my fascination with cars, and asked how I reconcile it with a strict training in academic art history, I reach for a gravure illustration of the 1967 Dino 206 SP's sixty-five-degree V6 and its three pairs of Weber 40 DCN/4 carburettors to make my point. I really do not know anything more beautiful.

No one makes a car with a carburettor today. Ford supplied the last Crown Victoria police cruiser (a personal favourite) fitted with a carburettor to the NYPD in 1991. In some export markets, Japanese pick-up manufacturers continued with carburettors until the mid-nineties and the very last car with a carburettor was, and no surprises here, a 2006 Lada. Meanwhile, Weber ended production in Italy in 1992 and Holley went bust in 2008.

Today, we have superior fuel-injection. Its advantages have long been known. In the Second World War, Rolls-Royce Merlin engines with carburettors would suffer from fuel starvation in extreme manoeuvres, while the injected BMW and Daimler engines of the Luftwaffe could maintain a continuous fuel flow even when pulling out of steep dives.

Of course, fuel injection is technically superior. Besides, mandatory catalytic converters require its precision. And OBD (on-board diagnostics) demands systems with a microchip-powered reporting ability. You don't get that sort of thing from imprecise, messy and temperamental carburettors. That's why they are so wonderful.

LUXURY STAINS

If, as I sometimes do, you spy on the conversations of motor industry professionals, you don't have to wait very long – usually no longer than the second glass of champagne – before someone goes glassy-eyed, froths at the mouth and starts emoting about the notions and meanings of 'luxury brand'. To discover it is a sacred quest, as potent as the continuing searches for the Kusanagi Sacred Treasures of Japan, El Dorado, the Nova Scotia Oak Island Money Pit and the missing Irish Crown Jewels. If only we could find it, we'd be rich.

But both 'luxury' and 'brand' are notoriously fugitive in definition. Luxury once suggested vice and depravity, so that's certainly stuff we can work with, but I prefer Coco Chanel's belief that it is not the opposite of poverty, but the opposite of vulgarity. I also rather like another definition from my old friend and mentor, Terence Conran, who said that any wine, even Algerian, served in a magnum becomes luxurious. As for 'brand', well we have brand managers with flip charts to tell us about that, but I'll settle for my own definition that it means the mixture of expectations and associations which successful products possess.

Certainly, cars have played a big part in popular understanding of luxury. In my own case, I still remember the ultra-violet glow of the instrument panel of my father's Jaguar Mark VIII. Even better, the walnut-veneered folding picnic tables in the backs of the front seats. These, never used as my mother inconveniently disapproved of picnics, seemed to anticipate pleasures yet to come and that, surely, is a luxurious experience. That they were redundant was of no importance: Stakhanovite functionalist directives play no part in luxurious all-sorts.

But the ultimate luxury car is, surely, the Citroën DS. Its prototype was the 'VGD', the Voiture de Grande Diffusion, which means, approximately, mass-market car. But the Citroën was nothing of the sort: it was divine, refined, sophisticated and entirely removed from the tedium of the everyday. More recently, Citroën decided to make 'DS' into a luxury brand. An element of the preparation for this was the opening of the DS World on Paris's rue François 1er, one of the most de-luxe addresses on earth.

While Citroën might sit more comfortably opposite Gap, DS World is opposite the couturier Balmain, evidence of ambition. But 'affectations can be dangerous!', as Gertrude Stein said of Isadora Duncan when her silk scarf wrapped itself around her Amilcar's wheels and she was theatrically flung to her death on the Promenade des Anglais, in Nice.

The 1955 DS presented a version of luxury with deep roots in French culture and business. It started with Michelin, when it decided to sell the idea of gastronomic road travel. Its guidebooks and its maps presented a persuasive vision of France as a network of fine restaurants and hotels, best connected by a comfortable car. It was just as the poet Stéphane Mallarmé said of Impressionism: '*peindre, non la chose, mais l'effet qu'elle produit*' – don't paint the thing, but the effect it produces. Michelin's Monsieur Bibendum, a cigar-smoking homunculus made of tyres, became the very symbol of felicity.

And when Michelin took over Citroën in a debt-for-equity swap, the tyre company had the influence to realise this vision in a car. So, the Citroën DS became the fulfilment of this dream of travel: a car designed to waft along autoroutes between a Michelin lunch and a Michelin dinner. The DS's extraordinary ride, moderated by its fantastic high-pressure hydraulics, combined with its sumptuous upholstery and a brightly lit cabin (which Roland Barthes described as the 'exaltation of glass') to provide the very optimum in automobile felicity.

On display in the rue François 1er are General de Gaulle's last DS and a superb Henri Chapron *décapotable*. The de Gaulle car is black with a light-grey interior, paying tribute to the austere tastes of the late President. The Chapron car is in a blink-making gold metallic with cherry red leather and glorious flashes of meretricious chrome excess around the wheels and lights, paying tribute to a louche Montmartre hostess. The two cars express the opposite poles of French culture, vernacular chic and grand luxe, just as perfectly as a Bic pen and a Charlotte Perriand chaise longue in pony skin.

Paris's DS World takes its place in an interesting short history of showrooms that do not sell (although new DS3, DS4 and DS5 cars are on display). Olivetti made its reputation in fifties New York by putting typewriters and office machines on display in a museum-like setting. Sony once did the same on Regent Street, in London. Each helped its parent accumulate valuable image capital … which was still being spent as Olivetti and Sony went into terminal decline.

They had forgotten what made them great. It wasn't 'luxury' and it wasn't 'brand'. It was 'product'.

INTERNAL CONFUSION

Apart from a gun, a classic car is the most analogue machine imaginable. Propulsion comes from an only-just controlled sequence of dirty explosions whose wasted energy is expressed as noise, heat and smoke.

Power is transmitted through a network of meshing cogs bathing in hot oil. The German word for 'gear' is *Zahnrad*, literally 'tooth-wheel', which nicely captures the almost feral nature of a mechanism that might bite you. Power continues to be sent through a long, spinning shaft before, absurdly, it turns ninety degrees via some larger tooth-wheels which humans elegiacally call a 'final drive'.

At the same time, the commander of this absurd mechanism will be sitting in a chair made from the tanned and dyed hide of a large mammal. The chair may have been hand-stitched by an artisan. Instrumentation will be flickering needles very likely set into a plank of polished wood or machine-turned metal. His or her hands will grip a 'wheel' which transmits crude manual inputs to mechanical manoeuvring system.

In comparison, the serene, clean, blameless and tranquil world of electrons and AI seems more rational. But I think it is the very irrationality of the suck-squeeze-bang-blow internal combustion engine, which literally drives creativity and gives character to the 'motor car'. There is no logical reason that a Juke and a Quattroporte coexist. But who wrote the rule that we must be logical? Form follows function? The hell with it! Form follows fiction.

For reasons related to the foregoing song of praise to the heat-engine 'motor car', I think it is fair to say that manufacturers of EVs have generally, shall we say, struggled to find a unique, or even simply relevant, design language for this new type, still less a pleasing one. Lenin's first Commissar of Business Efficiency was Nikolai Kondratieff, who had a theory that creativity came in huge, historic, cyclical waves, often inspired by new energy sources. And he was wrong. Lenin thought Kondratieff wrong too and sent him to the gulag.

Putting us in mind of both guns and the gulag was the fascinatingly awful baby-blue Kalashnikov CV-1 shown in Moscow, in 2018, a most unusual exercise in brand-extension by the people who brought you (and Hamas) the AK47. I still cannot decide if they were serious, but the design of this disturbing curiosity seems to have been inspired by the Fiat 124 of 1966 which, licensed to the USSR, became what we know as a Lada. Or perhaps the Moskvitch 412 was the source for this Togliatti Tesla. Maybe they have irony in the soul, but it is certain the Kalashnikov CV-1's designers lack aesthetic conviction. Or may possibly be vision impaired. But let us, perhaps, be prepared for a Purdey EV constructed of multiple layers of Damascene steel.

The artistic hesitation in EVs seems almost apologetic. And this is odd because, in historical perspective, electricity always seemed a better power source than petrol. Possibly the first viable EV was designed in 1884 by Thomas Parker, the inventor of a distillation of coal he branded as 'Coalite'. Parker was born in Coalbrookdale and thus had an affinity with industry. Indeed, his partner in the electric car design was a manufacturer of nails and horseshoes. Around 1900, when land speed records were held by electric cars, and what with Parker's reputation as 'the Edison of Europe', plus the success of the Flocken Elektrowagen and the Electrobat, any expert would have said the future was electric. Ironically, it was only when the electrical self-starter became a practical reality – obviating uncomfortable and inconvenient analogue cranking – that the feasibility of petrol engine cars became recognised.

The current enthusiasm for EVs can be traced to the early nineties when the California Air Resources Board (which makes the amusing acronym 'CARB') began agitating about pollution. The first-generation Toyota Prius of 1997 was one result. It was designed in California and looked like a plastic bath toy. Two years later, the GM EV1 featured faired-in rear arches, I think almost as a gesture of humility like an Amish woman's headdress and apron.

The second-generation Prius was altogether neater and more modern, but then we got to the fourth generation in 2015 and someone in Toyota's design centre said, 'we must lower the bar' and, with ruthless Japanese efficiency, they did. Maybe they do it to stigmatise Uber drivers, but the current Prius is surely in the running for best-of-show in the Ugly Dog Award. And how psychologically robust and sexually secure would you have to be to drive a Renault Twizy? Sure, Franz von Holzhausen's Teslas are very attractive, but that is because they look like motor cars.

And now I must fess up. In my part of London, to say 'I've got a Porsche' is as socially suicidal as saying 'Harvey Weinstein's my best friend'. So, I called my man at BMW and said, 'It's time to go electric'. And I chose an i3S because it is the first electric car to sketch the possibilities of a new design language and I find that compelling.

Yes, I know that John Heywood, Professor of Mechanical Engineering at MIT, said that in 2050 sixty per cent of cars will still be internal combustion. But expert predictions are always wrong. I have seen the future, and it hums. Or perhaps, pass me your gun.

SO LONG, SALON

I am fond of a good portent. You know. The usual. Thunder, lightning, dead dogs, strange births, dews of blood, stars with trains of fire, owls hooting in daylight, eclipses, eruptions and what Macbeth called 'dire combustion and confused events'.

The King of Scotland might have been describing the 2020 Geneva Salon International de l'Automobile. Except there was no such thing. The show did not go on. Images of the cavernous halls with half-built stands were as moving in their way as that part of Herculaneum where someone was incinerated while innocently eating his breakfast of garum and purple Falernian wine. No show. So, here was a portent, if ever there was one.

In the absence of media cleverly balancing plates of cheffy canapés and glasses of bad champagne at global reveals, new car launches that year had to be conducted online, a medium which suffers from viruses of its own.

It was striking (to us in portent-world) that the two most interesting new cars occupied the opposite poles of current technical possibilities and cultural demands: the all-electric Cinquecento and the all-Godalmighty Bentley Bacalar. That one is named after irrelevant piston displacement and the other after a Mexican resort speaks, I think, eloquently about the magnificent absurdity of the modern car's frame of reference.

Despite my *passione* for all things Italian, I will not be buying a battery-operated Fiat because my experience of electric cars is a deeply, deeply melancholy one: the practical problems are overwhelming, and the dynamic experience is not emotionally engaging. I don't want a mobility system – I want a way of life.

And nor will I be buying a Bentley. I don't doubt that the Bacalar is a superb machine and exciting in use, but to drive one in central London would be as embarrassing as having 'I have a severe-to-acute psycho-sexual problem and I am also a shameless environmental terrorist' tattooed on my forehead. It would also be impossible to park. So, no deal, really. Thus, the Fiat and the Bentley describe the critical absurdity of the contemporary motor trade.

From the harrowing perspective of now, we can see that the story of the motor car was – actually and metaphorically – about the 'conquest of space', the title of a ludicrous 1955 sci-fi movie. This was why, circa 1957, Harley Earl told his GM designers to make cars look like missiles. This was why Europeans fantasised about 'grand touring' cars, vehicles to take them on romantic (and imaginary) journeys.

And with the parables of escape came the iconography of prestige, of sex and success. In a world where Greta's parents had not been born and concepts of consumption were not morally tested, you get a '59 Cadillac. Say what you like about this car, but a world in which it was possible was a world full of uninhibited possibilities. Or put it this way: the poverty of contemporary car culture is such that my grandchildren will surely not be writing whimsical and nostalgic essays about the Nissan Leaf.

Cars no longer influence pop music. Nor any form of pop culture. When did you last see a film in which an automobile was a proxy for a personality? The best TV I have ever watched was HBO's *Succession* and the characters therein spend what little time they do in cars sitting in the back of an S-Class or an Escalade. I find that very revealing: they have no interest in being in command. They use cars passively. The paradise of personal mobility is coming to an end.

Whether a car or a building, great designs are always related to their environment and circumstances. Where else would you want to drive a notchback '65 289 Mustang than the Pacific Coast Highway? A Mini Cooper S has its scriptural home on the King's Road. A 2CV was built for rolling down the old Route Nationale between Avallon and Beaune, preferably after a Cadillac-sized lunch.

Follow this reasoning and you wonder what a modern car might actually be. One designed for me would have to cope with seven sets of traffic lights within a mile of my front door, London kerbs, blanket 20mph limits and a concrete bunker at Waitrose, where I can't get the signal I need to fire the app to charge the wretched thing.

So, not for the first time, I take refuge in books. Sydney Charles Houghton – 'Sammy' – Davis won Le Mans in 1927. In 1952, he published *Car Driving as an Art*. I'd prefer to call it a high craft, but you see his point. Or another favourite, Hugh Merrick's *The Great Motor Highways of the Alps*, published in 1958. The author's photograph shows a beaming chap with a pipe, ready for another adventure on the Val Tremola windings of St Gotthard or the Furka hairpins.

I enjoy these dreams more than I enjoy the confused events of a shuttered and tarped Salon de l'Automobile. Marcel Proust, the greatest poet of loss (who actually found cars very interesting) remarked 'the true paradises are the paradises we have lost'. Q: Whither the motor show? A: '*Fine del mondo. Punto*' a friend wrote to me.

GREY

Lockheed had its Skunk Works where Clarence – 'Kelly' – Johnson, the Harley Earl of aerospace, worked on secret projects, including the twin boom XP-38 that the RAF flew as the 'Lightning' and, later, the U2 spyplane and the SR-71 'Blackbird'. At 2,193mph, the latter holds the all-time speed record for an air-breathing craft and once travelled from New York to London in ('Geez, we haven't got time to stop for a pee') I hour 54 minutes and 56 seconds.

I met a Blackbird captain once at Mildenhall AFB and asked him what it felt like to experience an 'unstart', the USAF's splendid euphemism for a lurid stall in one of its psychotic ramjets. He said, smiling, that the resultant violent asymmetric thrust that computerised flight controls took dreadful, long seconds to detect, 'Just watered the shit right out of your eyes'.

Be that as it may, I know of somewhere in the car industry at least as obsessive in its commitment to wacko R&D. This is Audi's top-secret Abteilung für Grau-forschung (Office of Grey Research), a sister to its equally clandestine Büro von Kunststoff-Studien (Bureau of Plastic Studies). In the matter of exploiting the aesthetic limits of the greyscale and the tactile extremes of the touch-of-plastic sensation, Audi has no equals.

Lockheed's Skunk Works was named after a moonshine factory and, clearly, the Germans have been imbibing something intoxicating too. A recent visitor to the site heard from behind the airlocked doors muffled cries of '*Hilfe! Ich habe meine Farbmuster verloren!*' (Help I have lost my colour swatches!). People were arguing about psychometric nuances between grey metallic and grey flat, between clear lacquer topcoat and matt finish. He saw new editions of Wolfgang von Goethe's *Farbenlehre* (Colour Theory) being wheeled in on trolleys. They were asphalt-coloured Alcantara volumes with eye-wateringly reductivist typography by disciples of the late Otl Aicher (the graphic designer whose most frivolous job was the Lufthansa logo).

Personally, never mind head-butting sans serif, I am fond of grey and think the paradoxically achromatic colour is one of the most subtle, handsome and interesting ways to paint a car. True, grey is the colour of ash, of death, of disease and of civil servants or politicians, but there is exceptional dignity and expressive value in it as well. It is also the colour of brains. So far from signifying bland reticence, the choice of a grey colour scheme bespeaks bold confidence in the consumer. It's significant that Audi's most steroidally inflated fast cars look so much more impressive in solid grey than in supposedly eye-catching and exciting bright colours. Today, more humble makers do nice greys as well. I am betting Mini's new grey becomes one of its most popular colours this year and there is no easier way of looking at a Skoda Fabia than at one painted like an Austro-Hungarian siege howitzer. And if anyone is tut-tutting and thinking of Rosso Corsa, never forget that Italy was assigned black as an international racing colour before it adopted red. And Abarths raced in a very light grey that survives as a lovely option on the new Cinquecento 595.

Certainly, now that high-definition colour-printing and faithful colour on smartphone screens are commonplace, black, white and, most especially, grey achieve sophisticated distinction. Future cultural historians will, surely, see some special significance in the fact that, in 2016, after a century of painstaking research in colour photography, Leica introduced the 'M Monochrom' camera, which does absolutely anything you can imagine pictorially except capture images in anything other than black and white. Here, if it were actually needed, is proof of Thorstein Veblen's theories about conspicuous consumption and the fickleness of consumers.

But this may yet be a critical moment in the history of car chromatics: the signs are mixed. It is interesting that, in the nineties, where sumptuary laws allowed, tennis players would wear shirts with absurd different-coloured sleeves. The anarchy continues. While once national teams were happy to wear single-colour strip, today they insist on contrasting socks and shirts with two, three or more different-coloured panels. This may explain England's disasters.

I am put in mind of the notorious Ford Edsel which, in 1958, was available in tricolour editions: before me is a picture of one in Sunset Coral, Snow White and Jet Black. Perhaps this, more than the supposed resemblance of its grille to a chromium-plated take on female genitalia, was a reason for its catastrophic failure.

It is unarguable that colour reveals psychological states. Goethe tells us that *Violett* (violet) is *unnoethig* or 'unnecessary' and that *Blau* (blue) is 'common'. Now that black, white and silver indicate only the existence of a minicab, they may also be untouchable. Even if it has strong associations with the military and with boredom, grey, I think, is beyond fashion.

Polychromists may argue differently. The original Benz, of 1886, the machine that started this all, in farts and belches of grey smoke in Mannheim, had a green chassis with a conspicuous red engine. The acid Arancio on the original Miura and the blinking Verde Pistachio on later cars helped make Lamborghini's reputation for sensationalism. But the greatest car ever made was manufactured in grey alone for its first decade. What was this car? The Citroën 2CV. I rest my case and shut my eyes.

TOM WOLFE

The Best American Writer is – or was – Tom Wolfe. Yes! I can call him Tom because I'm pleased to say I knew him a bit. Anyway, Tom, for me, is the greatest writer not just of American, but of English. He is better than literature.

His incubator, *Esquire* magazine, said his work was 'so thrilling, and funny, fresh and new that it changed everything'. That's an understatement. What happened was that Tom's vivacious 'New Journalism' translated low culture into high art, making conventional novelists seem flat-footed and dumb. And no one sensed better what cars meant in the late twentieth century: that first sentence above is a take on the title of one of Tom's first and greatest stories: 'The Last American Hero is Junior Johnson. Yes!', Johnson being the great NASCAR racer.

Wolfe's Junior Johnson article was published in the March 1965 edition of *Esquire*, whose cover was George Lois's famous image of a woman wet-shaving with lots of foam, so this really was an altitudinous high-point in the already glorious history of American magazine journalism.

Tom, a college boy, went to his first NASCAR race at Wilkesboro, NC, wearing a tailored green tweed suit and a dashing Borsalino fedora. This was a rare misreading of local customs and, altogether, an unsuccessful attempt to go native. Still, the alienation achieved may have enhanced his sense of detached wonder: at Wilkesboro, Tom witnessed and was impressed by Junior Johnson's innovation of 'drafting', or what more effete Europeans call 'slipstreaming'.

And if the sight of a bellowing '63 Chevy tailgating a '63 Holman & Moody Ford at 175mph was not interesting enough, Tom was very taken by Johnson's character. The son of a famous bootlegger, who had spent a third of his life in jail, notably for the biggest booze bust in US history, Junior Johnson was, additionally, an Appalachian dirt-racer, chicken farmer, 'coon hunter and all-round good ole boy of robust tastes. After winning fifty NASCAR races in exceptional style, Johnson retired to run his own team where he managed Cale Yarborough and Darrell Waltrip. President Reagan pardoned him in 1986 for a bootlegging offence from thirty years earlier and, newly confident, Johnson began selling his own-brand Midnight Moon Moonshine in 2007.

But this was not Tom's first adventure with cars. A visit to Los Angeles had introduced him to hot-rods. Wolfe, a white-shoe Southern aristocrat with a PhD from Yale, was amazed at the scene. His report appeared in *Esquire* in November 1963 with the epic title: 'There Goes (Varoom! Varoom!) That

Kandy-Kolored (Thphhhhhh!) Tangerine-Flake Streamline Baby (Rahghhhh!) Around the Bend (Brummmmmmmmmmmmmmmmmmm . . .)'. Until this point entirely innocent of automobile culture, Tom became fascinated by the noise, glamour and mesmerising symbolism, noting in a now famous formulation that, to Californian kids, cars meant 'freedom, style, sex, power, motion, colour, everything'. Another understatement.

Later, came Tom's veneration of the road trip. Inspired by Jack Kerouac's 1957 classic *On The Road,* an odyssey with a V8*,* the writer Ken – *One Flew Over the Cuckoo's Nest* – Kesey had bought a 1939 International Harvester school bus (which he christened 'Further') and decided to drive from Los Angeles to New York with a group of intoxicated, tranquilised and hallucinating friends called 'The Merry Pranksters'. Tom turned this into potent mythology.

Before departure, they customised Further with a rooftop turret made out of an old washing-machine drum, a shelf out back to carry a motorbike and an electrical generator, as well as a most distinctive psychedelic paint job. Fuelled by beer, LSD, Benzedrine and marijuana, the bus set off from La Honda, California, on the 17th of June 1969.

They did not quite know the way to San Jose, the first stop, and, with Beat Generation cowboy Neal Cassady at the wheel, covered only forty miles in the first twenty-four hours. Twelve days later, in some disarray, The Merry Pranksters and their bus arrived in New York to be received by Kerouac himself with his chum, the poet, Allen Ginsberg. The bus made a final trip to Woodstock and died in an Oregon field.

Even if he did not participate, the trip was described with hypnotising accuracy by Tom in his essay 'The Electric Kool-Aid Acid Test' which, with Hunter S Thompson's 1971 *Fear and Loathing in Las Vegas*, is the best account of two-lane blacktop camaraderie. Or: 'You're either on the bus or off the bus', in words Tom put in Kesey's mouth.

True, Ginsberg said, 'I saw the best minds of my generation destroyed by madness'. But the exhilarating thing is, the madness of draughting seven-litre wallowing Detroit barges at improbable speeds on rebel yell ovals did Junior Johnson no harm whatsoever: like Tom, he lived into his late eighties. Given his habits, Kesey too endured.

Tom Wolfe saw romance in a NASCAR Chevy and also in a '39 International Harvester bus: Southern redneck heroics on the one hand, West Coast counterculture nomads on the other. There's America's breadth expressed in vehicles, and no one understood that better than Tom. Mother dog! Yes! Varoom! Varoom! His style is addictive.

WHEELS

Cars are defined by their wheels, obviously. Without wheels, a car would be a stationary metal hut whose engine existed only to power the air-con, and the essential distinction between mobile and static architecture would have disappeared. So, designers are taught to take wheels very seriously. It's a part of studio tradecraft to know the optimum ratios between wheel height and beltline, for example. It affects the whole aspect of the car.

These ratios are numbers that every designer knows. More subjectively, we enjoy looking at cars whose wheels and tyres seem to fit the arch in a way that suggests swollen muscularity straining at a T-shirt's sleeves. If wheels look lost in their arches, a car seems weedy and maladroit. This was the only error spoiling the otherwise consummate masterpiece that was the NSU Ro80. Or, compare a 427 Cobra with a Mini Metro (the latter compromised by having to use old narrow Mini subframes in a new wider body, the former being influenced by the great gobs of V8 torque urgently needing management-by-tyre).

And in the contemporary marketplace, designers also know how a car's precise position in the social hierarchy may be determined by the wheels alone. Detroit used to call this status apartheid 'series differentials', meaning how different interior trims indicated what money you had spent on your Pontiac. Stuttgart, Munich and Ingolstadt now outperform Detroit in this psychological intimidation as effortlessly as they do in dynamics. If you buy a cheap Audi A4, you are forever stigmatised by its small wheels with apologetic detailing. It's not that the overwrought and enormous wheels on a top-of-the-range car look like a gross addition, it's more that the simple wheels look like a humiliating deduction. It's that cruel.

Interestingly, the English and the Americans have played a big part in the history of the car's wheel. One hundred and forty years after Humphry Davy identified magnesium as an element, Ted Halibrand starting sand-casting mag wheels in California. With his forge, Halibrand put the word 'mags' into the language. His influences came from aerospace (Lockheed was nearby in Burbank) and his market was the new clique of hot-rodders who used the liberations of the GI Bill to test human tolerance to pearlescent paint and high acceleration.

Halibrand's distinctive wheels, with their very technical aspect and fine ribbing, appeared on the Thorne Engineering Special that won the Indianapolis 500 of 1946. Soon, another wheel with kidney-shaped ventilation slots became a customiser's favourite and his signature item. Later, a finer-pattern Halibrand wheel was used in most Briggs Cunningham specials and was standard fit on factory Ford GT40s and Carroll Shelby's Daytona Cobras. With their wildly exaggerated three-ear spinners, these Halibrands became immediate identifiers of some of the greatest sports cars of the sixties.

The classic threaded wire wheels which cast magnesium eventually replaced found their scriptural home in Milan at Ruote Borrani, founded in 1922. But Borrani's design for a lockable wheel on a splined shaft was based on a British bicycle patent. Indeed, Borrani emerged from an earlier entity called Rudge-Whitworth Milano. So, with its origins in Coventry, the classic seventy-two-spoke Borrani wire wheel was the standard fit on Ferraris made between 1946 and 1966.

Porsche too was distinguished by its wheels. Before adopting discs, Porsche had a unique system where, to avoid duplicating components, the studs carrying the wheel were integral with the brake drum, meaning that the 'wheel' was in fact just a steel rim, an annular device, carrying the tyre. This made, say, a 550 RSK, look exceptionally technological underneath the arches, although the great Porsche expert, Karl Ludvigsen, tells me the design was copied from a Ford truck seen by Porsche when visiting America. So, Midlands bicycles influenced Ferrari and Midwest trucks influenced Porsche.

Again, in the sixties, the classic 911 was defined by its alloy wheels designed by Heinrich Klie for the Otto Fuchs Metallwerke. Other wonderful examples include the laboratory-precise Dunlop ventilated wheels on the Jaguar D-types, and the idiosyncratic 'Wobbly Web', designed by Lotus's Gilbert McIntosh and inspired by a nineteenth-century pulley, but made famous by the egregious and sharp-elbowed Colin Chapman.

On road cars today, size alone seems the chief determinant of wheel design. And I think this is a very German thing since it lends a sense of aggression – rather than delicacy – to a car. And this is enhanced by the absurdity of extremely low-profile tyres. Certainly, less flexible low-profile tyres behave in ways which can enhance performance, but if they were dynamically essential, Formula One cars would have them too. Instead, Formula One cars have notably high-profile tyres because the compliance of the sidewall adds a measure of 'comfort' which the nearly solid suspension does not allow. Design is always about a compromise.

Certainly, design students today are still busy drawing concepts with wheels that look like rollers smeared in rubber rather than discs wearing tyres. But pausing recently to stare mournfully at a photograph of the crashed Facel Vega in which Albert Camus died, it's remarkable how vast were the tyres (at least until one of them exploded). It occurred to me how ridiculous an Audi A8 would look with balloon-like tyres on small wheels. And that got me thinking ... maybe it is time for another aesthetic direction in wheel design.

FACTORIES

Ford's Highland Park assembly line, 1913

Karl Marx noted that 'nature makes no machines'. But men
do. Albert Kahn, America's master factory designer, built
Ford's second assembly plant on Woodward Avenue, in
Detroit, between 1909 and 1914. Although Henry Ford
himself was perhaps was not the engineer, the moving
assembly line was a clear demonstration of his principles:
division of labour, process optimisation, time-and-motion,
cost cutting, and integrated supply chain. The brutality
of the system was satirised by John Dos Passos, but its
efficiencies meant the cost of a Model T was halved
between 1910 and 1917. Ford may have brutalised his
workers, but in compensation paid them three times
the Detroit average.

FIAT's Lingotto *pista*, 1930s

The greatest contribution FIAT made to Turin's architecture
was the astonishing Lingotto factory on via Nizza. This was
a uniquely ambitious project, begun in 1916, and the sole
building of note designed by a naval architect from the
Ansaldo conglomerate, called Giacomo Mattè-Trucco
(1869-1934). Inspired by the ravings of the Futurists and
the unrealised architectural visions of Antonio Sant'Elia,
Mattè-Trucco created a vast 400,000-square-metre
concrete factory. At each end, helicoidal ramps rise five
storeys to the rooftop *pista*. Le Corbusier was so
impressed by this expression of the new spirit in building
design, he cited Lingotto in *Vers une architecture* (1923).

Volkswagen's Wolfsburg factory, 1951

This is where the sinister KdF-Wagen became the lovable
VW Beetle. The cornerstone of the Stadt des KdF-Wagens
bei Fallersleben was laid on 26th May 1938: a ceremony
conducted with maximum Nazi pomp. Military Kübelwagen
and Schwimmwagen were also made here. Porsche had,
indeed, conceded that his Volkswagen might well have
military applications. In 1945, the British Occupation Forces
renamed Fallersleben 'Wolfsburg'. The British, advised
by experts from Humber, questioned the Volkswagen's
commercial prospects. Soon, however, the plant became
symbolic of Germany's remarkable *Wirtschaftswunder*.
The Volkswagen Group now comprises, not only VW, but
Audi, SEAT, Skoda, Bentley and Lamborghini. Humber is
no more.

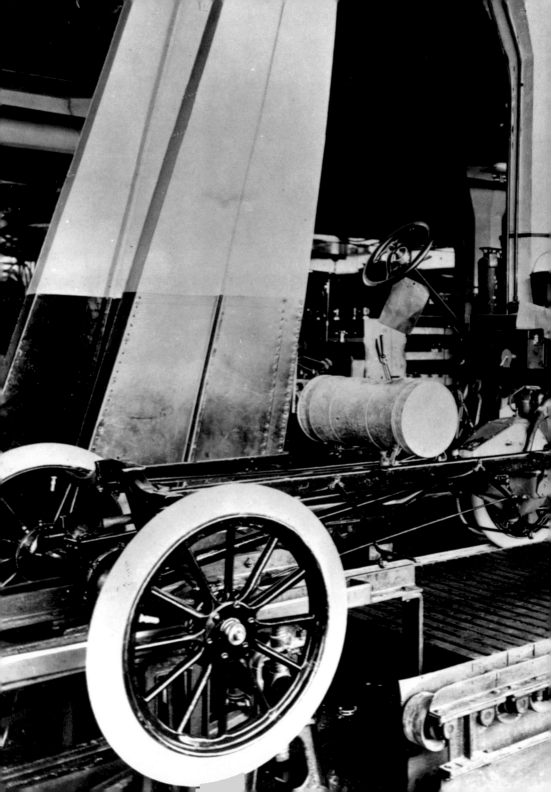

'62 CAPRI

A drum I have been banging for a very long time beats out the message that cars can be considered as art. In my very first book, published more than forty years ago, I mentioned Michelangelo and Raymond Loewy – authors, respectively of the St Peter's *Pietà* and the Studebaker Avanti – in the same paragraph. Where it was not ignored as the rambling of a hallucinating maniac, in 1979 this assertion caused spittle-flecked outrage.

And how right I was. It's not that I think cars *are* art, just that in their sculptural presence, their ability to express and define collective yearnings, their claim on beauty, their narrative density, their emotional appeal, they have usurped the traditional role of art. Anyone who has seen the boring, posturing absurdities of the Turner Prize will agree. So do the organisers of Art Basel Miami, the planet's biggest art fair. In 2018, an event called Grand Basel was inaugurated. This was the formal beginning of a process that will, commercially speaking, make collectible cars as artistically acceptable as Warhol and Koons.

The focus of Grand Basel was an ambitious 1953 concept called Linea Diamante, by Gio Ponti, architect of Milan's Torre Pirelli. It was recreated here by Roberto Giolito, who, incidentally, was designer of the current Fiat 500, but is now at Fiat's Centro Storico, a guardian of its culture. I was one of several people invited to nominate production cars to be put in oversized frames and be cast as 'art'.

Given a free choice, I chose not something by Figoni et Falaschi, but a '62 Ford Capri. You will want to know why. One reason is my fascination with Henry Ford's remark that 'you can read any object like a book'. And that's true. Anything that is made betrays the beliefs and preoccupations, doubts, fears and desires of the people who made it. Just like great paintings.

The idea of what became the Ford Capri entered the mind of Ford of Britain's designers in 1956. This was an anguished year: the shaming calamity of Suez marked the final end of the fading British Imperium. Pride was damaged. 'Project Sunbird' was begun to give demoralised designers something to dream about.

And they dreamt about America, because Britain always had a cadet role in the 'special relationship' and the US offered a vista of attractive possibilities unavailable at home. The dreamers were led by Colin Neale, who had absorbed the highly competitive design culture that eventually created the '61 Lincoln and '62 Thunderbird. His experience of Dearborn was so competitive, he called it 'the stiletto studio', referring to the needle-thin dagger that was an assassin's favourite weapon in the Renaissance.

But the Capri was also a product of management fiat, a command from on high. This flamboyant, pillarless coupé was to be a 'personal car', with all the psychological subtleties that implies. It was to be a car you could take to the golf club, which is to say socially superior to most proletarian Fords. Hilariously, it was also to be a 'co-respondent's' car, referring to the legal term for the third party in a divorce. Such was the almost erotically seductive power of the idea. Such was the dream.

The result was one of the strangest mass-produced cars ever created. Neale (who also drew the very successful '59 Anglia) took styling cues from the Ford Galaxie and the Fairlane Skyliner. The huge rear deck is reminiscent of the '61 Lincoln. He said the Capri was 'sculpture in sheet metal', an expression he might have borrowed from Philip Johnson at New York's Museum of Modern Art. And the Capri's formal complexity made it, like sculpture, ruinously expensive to manufacture: Ford had to subcontract to Pressed Steel Fisher.

But the designers and marketeers dreamt of 'the Continent' too. The Capri's contemporary Ford Cortina was named after a Dolomite ski resort. The Cortina sales brochure used the graphic motif of passport stamps to suggest cosmopolitan sophistication. The new package holidays now even made exotic Capri accessible, as if the consumer could, via a car, access a Tyrrhenian island as readily as he once accessed Oxford, Cambridge or Westminster.

And personal cars came in a range of colours bizarrely at odds with the English climate: a limoncello yellow and pistachio green, an ivory caffe latte. It was the first popular car to wear a 'GT' badge: the eighteenth-century Grand Tour having been the historical origins of Anglo-Continental voyeurism. And while there was nothing very much technically distinguished about the Capri, it was the first popular British car to use a Weber carburettor, establishing a vicarious connection to Lancia, Ferrari and Maserati.

Henry Ford II liked the Capri so much that he gave one to his daughter Charlotte. But that affection was not shared by the general public: it was a sales calamity. Dreams are not reality. A mere 19,421 were made (and very few survive), making it one of the rarest Fords ever; 19,421 is more than Koons, but fewer than Warhol.

Still, the '62 Capri is now officially a work of art. I know – it's been framed.

GENIUS LOCI

I have been blipping, shuddering and double-de-clutching a modern classic Fiat Cinquecento, Dante Giacosa's 1957 miniature masterpiece, around the mountains of north-eastern Sicily. The mountains were green, the sky was blue, and the car was that iffy period colour known as Positano Yellow.

This raised the spirits mightily and reconfirmed two long-held beliefs of mine. One, that the enjoyment to be had from any car's performance has nothing to do with absolute speed, the calculus of acceleration, or the generation of eye-popping centrifugal force, but has everything to do with the relationship between input and response. Aesthetes, you see, are sensitive to balance and proportion (as well as colour).

Of course, the little Fiat is remarkably slow, and the brakes feel unwilling to meet even the lowest expectations in the matter of retardation. But the steering is direct and true, weight distribution interesting, while the limitations of power mean that you can drive near the car's limits and perhaps beyond your own on a continuous basis. This means that, technically speaking, you cannot wipe the smile off your face.

Two, that certain cars and places have an almost mystical bond. In his wonderfully sardonic novel, *Up in the Air* (made into a wonderfully sardonic movie starring George Clooney), author Walter Kirn says there is 'no pleasure more reliable than consuming a great American brand against the backdrop featured in the advertising'. And in proof he cites a Ford F-150 on a dirt road or a Coke on the beach in Malibu.

It is the same with the Cinquecento in Sicily. This car would be preposterous on Park Avenue and scary on the M25, but on the 1,380-foot descent from Forza d'Agro (Coppola's *Godfather* village) to the sea, you feel connected: all the muddle and confusion and conflict of the world disappears and everything appears correct and aligned. Forget about the tentative orange plastic netting replacing barriers lost to landslides at the blindest of corners, 'you're in the right place, you're running with the right forces, and if the wind should howl tomorrow, let it', as Kirn puts it.

This got me started on other cars and their special relationship to places. It's a two-way thing. On Park Avenue, for example, you would only ever want to be in the back seat of a Ford Crown Victoria or a Chevrolet Caprice yellow cab. On the other hand, you would not want to be in one anywhere else. Still, Manhattan itself will be diminished with their passing: if there are gods in New

York, they caused the Crown Vic and the Chevy to exist. These gods will not be pleased by the Nissan replacement.

On the Upper Corniche, high above the Côte d'Azur, I only ever think of a 1953 metallic blue Sunbeam-Talbot Alpine Mark I (with cream upholstery). This was the car in which the tempestuous Grace Kelly scared the urbane Cary Grant in Hitchcock's *To Catch a Thief*. There is meaning here: Kelly was the beautiful daughter of an Irish family whose genes were remodelled by Philadelphia money. And the Alpine went through a similar transformation: a dull Sunbeam 90 saloon was turned into a dashing roadster by Hartwell's of Bournemouth, with some assistance from that master of schlock in a cloud of eau de cologne, Raymond Loewy. Kelly and the car were, you see, each highly contrived American designs.

The Beachboys' 1955 Thunderbird? Venice Beach, CA 90291 for sure. A '64 Mustang? This car belongs in only one place: Le Touquet, where it starred in Claude Lelouch's maudlin, but beautiful *A Man and a Woman*. Somehow, the dirty butch V8 and coruscating metallescent Detroit kitsch become fetching, delicate and wistful in the elegant northern French *campagne*. I can hear the film's annoying theme tune, with the 1-5-4-2-6-3-7-8 firing order as a base note, even as I write.

But consider then Hollywood's most haunting use of a car as symbol of yearning: Dustin Hoffman in the bright red 1966 Alfa-Romeo Duetto, the *coda di seppia* original, rushing to disturb the ill-starred wedding that is the climax of *The Graduate*. He is crossing the Bay Bridge from Yerba Buena Island to the Embarcadero. This is the triple distillation of the automobile as a romantic accessory and it is a journey I long to replicate. The Duetto never was a very good car, and I have no interest in driving one anywhere at any other time, but in San Francisco crossing that bridge with the roof down it would be near enough a religious experience.

There is a sombre and incongruous side to this ineffable car and place relationship. The magnificent Facel Vega HK500 will not be remembered on the Champs-Élysées, but by the tree near the bleak village of Villeblevin, south of Paris, which Michel Gallimard's example hit in 1960, killing his famous passenger, Albert Camus. Twenty-seven years after the Alpine, Grace Kelly died on the same Corniche when her Rover P6 left the road.

There really is no doubt that cars have added meaning to romance … and to tragedy.

MICHELOTTI

Who was Giovanni Michelotti? Of course, the name is familiar. And so too are the cars, which include the Triumphs Herald and Spitfire: unique shapes and likeable machines no matter what their technical shortcomings might be.

The familiarity of the cars is not surprising, because in his short life, Michelotti – always known as 'Micho' – drew more than 1,200 of them. But here is the enigma of a man so varied and prolific he styled both the two-stroke fibreglass Frisky, of 1957, and the ohmygod Ferrari 340 Mexico, of 1952, yet remains almost unknown as an individual.

I found an article from the *News Chronicle*, of 1957, boldly titled 'He Gives Cars a New Look', referring to that engagement with Triumph that, via the Herald, ultimately led to the Stag. The journalist had visited him in his Turin studio-home on the sixth floor of an apartment building at 35 Corso Francia, not far from the modern art gallery. The thirty-six-year-old Micho, described as reserved, even taciturn, was furiously at work, chain-smoking, shuffling French curves and pencils, on a *tavolo da disegno* 'twice the length of a family car'. Let's assume that's a European family, not an American one.

He described himself as a *creatore* and explained that he had drawn obsessively from childhood. 'And when I was not drawing, I dodged lessons and went dancing with my friends.' His only diversions from eighteen-hour days were two children and landscape painting in oils. His sole declared vice, besides dancing, was *gelato* doused in brandy, a vice he indulged often. He drove a grey Fiat saloon with a clandestinely tweaked engine. At the time of this interview, Micho had never been to Venice and to Rome only once … and that a mere six months before.

So, what we have here is a provincial prodigy of astonishing energy and commitment. Micho was described by Luca Cifferi as a 'pencil soloist' and indeed he was a virtuoso draughtsman. If his drawings sometimes seem a little unfinished, even crude, the personal handwriting is always idiosyncratically distinctive and attractively energetic. His work-rate was prodigious. His son Edgardo told Ian – Phantom – Cameron: 'He never had time to crap since in the same time he could knock off another sketch and get paid for it'. Cameron added darkly: 'which is what a lot of his work looked like!'

But Micho was a soloist in another way. As soon as it was possible to do so, he left the *carrozzeria* system of Turin with its feudal relationships, its hierarchies, its pride, vanities and its jealousies, to become a lone operator: perhaps the very first independent stylist in Europe.

And now that the *carrozzeria* tradition is dead, we can perhaps acquire by looking at Michelotti a better and clearer idea of questions of provenance and authorship in the confused world of Italian design.

At fourteen, an uncle who made *bozzetti*, the full-size wooden maquettes that coachbuilders used as formers for bashing metal, got him a job with Stabilimenti Farina, the manufacturing part of that family business. In 1949, aged twenty-eight, Micho became independent, but worked closely with Vignale, Ferrari's preferred *carrozzeria* in the period after Touring and before Pininfarina. Significantly, he spurned every job offer and there were many, insisting on his independence. He styled an entire car and two facelifts *every* single week. His facility became legend. The Frisky was the work of twenty-four hours. And so too was the Triumph Herald. Perhaps that showed. '*C'est facile!*' he said to Triumph engineer, Harry Webster, in the French they used as a common language. His work on the Herald cost Standard-Triumph £10,000.

In this way, more than thirty cars at the '57 Salone in Turin were by Micho's hand, although none carried his autograph. Why? Because even as an independent, he was still in the shadow of the *carrozzeria*. And this was how it remained until the coachbuilders became defunct: Franco Scaglione, Leonardo Fioravanti, Paolo Martin, Lorenzo Ramaciotti, Aldo Brovarone, Giovanni Savonuzzi and even Giugiaro were the designers. But Pininfarina, Ghia and Bertone took precedence as the brands.

When BMW seemed directionless, Micho drew the sweet little 700, in 1959. That aesthetic developed into the Neue Klasse of 1961 which, until Chris – Disruption – Bangle, defined BMW style. But, of course, Wilhelm Hofmeister took the credit. Micho also introduced the Japanese to European style and expertise, the first designer to do so. The Prince Skyline Sport shown at Turin '60 went into production in 1962. In 1964, he reversed the Italo-Japanese trade and brought Tateo Uchida to work with him in Turin. Another first.

So, the answer to the enigma? I asked my friend Stefano Pasini, a true connoisseur of the motor car, for his opinion:

'I have always admired the way Micho understood, and translated into metal and glass, the intrinsic character of the marques he was asked to cooperate with. So, his BMWs were lithe and innovative and unmistakably German, while his Triumph TR4, though miles apart from the TR3 it superseded, is still the epitome of British sports cars. How did he do that? Beats me.'

Well. I have a clue. Looking at the photograph of that figure crouched over his *tavolo da disegno*, restlessly drawing, ignoring reason and sense, helplessly creative, sometimes crass, sometimes wonderful, you can see a true artist at work. Micho died in 1980, aged fifty-nine.

NEUNELFER

I think the eighties will always, even if only ironically, be known as the Design Decade. To me, at any rate, they certainly were. I rode shotgun on a sleek, high-speed, well-designed bandwagon driven by Sir Terence Conran, the man who (I think quite correctly) believed that universal happiness could be sourced in a better salad bowl. Terence owned thirteen 911s.

The decade began with an experiment of ours called the Boilerhouse Project, in the Victoria & Albert Museum, where an exhibition about megabrand Coca-Cola outraged art-world traditionalists but became one of the V&A's most successful-ever shows. It ended with the opening of our Design Museum. It's not for me to say whether the popularity of the Boilerhouse and the Design Museum were a symptom or a cause of the pervasive consumer mania of the age, but the eighties made 'design', hitherto discussed only by consenting adults in private, the familiar subject it is today.

And the eighties was the decade when Porsche changed. Dr Porsche's original inspiration was: 'I could not find the sports car I wanted, so I built it myself'. That was 1948. For over thirty years, superb cars of the greatest technical and aesthetic purity were the result of this solipsistic brief-to-self. To my mind and eye, the last authentic expression of the Porsche idea was the 911 SC, which ended production in 1983.

During the eighties, design ceased to be a description of an activity, of what somebody does. Instead, it became a commodity to be acquired and I realise I'm partly to blame. Certain preferred objects were held to possess this quality, while others did not. This led to a lot of competitively absurd nonsense and austere Porsche was not immune from contamination.

By the end of the eighties, the original Porsche Type 901 had ceased to be a mere sports car and had become the genetic source of a valuable brand. This brand has ever since been exploited with German thoroughness and American cynicism. The Cayenne? Try in your imagination explaining the rationale of this hideous, inelegant, overweight atrocity to Dr Porsche, an engineer who counted the *Windkanal-Pionier*, Paul Jaray, among his collaborators.

I met Ferdinand Alexander Porsche, the Ur-Dr's grandson, always known as 'Butzi', in 1981. The location was Zell am See in Austria, tribal seat of the Porsche family. Apprentices were sent here to a Porsche-badged holiday camp which, even to the insensitive, carried a slight aroma of the Third Reich.

Ferdinand Alexander had drawn that first 901, so has an undisputed place in the Automobile Hall of Fame, but he appeared to have been professionally banished to Austria because of a family dispute whose source was never made quite clear. Here he was cultivating ideas for Porsche Design, the independent studio he had founded in Stuttgart, in 1972. The Porsche car company only took over Porsche Design in 2003, ending an ambiguity about Porsche Design's authenticity which some say Butzi deliberately exploited.

Anyway, if you wanted a symbol of eighties high-concept, big-ticket consumerist hedonism, you could not do better than Porsche Design aviator sunglasses with clip-in, clip-out faded-tint apricot-coloured lenses, which made you look like a Bulgarian paedophile. I bought mine at New York's Museum of Modern Art shop, in 1979. Porsche Design now sells, one might say 'over designs', briefcases, pens and ski boots. This brand stretch reached its culmination in 2017 when the sixty-storey Porsche Design Tower was inaugurated at Sunny Isles Beach, Florida. The headline feature here was robotic parking, with cars taking lifts to aerial garages adjacent to the apartments. 'Selling-out like cheap gasoline' was one headline, the writer unaware of the double entendre.

Decades to not have GPS accurate start-points. As everyone knows, the sixties only began in 1963 with The Beatles' first LP and the almost simultaneous discovery of sexual intercourse. Porsche's eighties began in 1975 with the 924, a very fine shape by Harm Lagaay, but a mediocre car by Porsche standards. It is said that Volkswagen, in a gesture Freud might have understood, violently rejected it as a Scirocco proposal. Terence Conran's sister bought one of the very first. In 1977 the 924 was complemented by Wolfgang Mobius's superbly odd 928. These two front-engine cars were the beginning of Porsche's dilution from a single potent idea to the sprawling and incoherent proposition it is today.

And it was also in 1981 that I met the designer Anatole Lapine, who had joined Porsche from General Motors twelve years before. In Detroit, Lapine had worked under Bill Mitchell on the fabulous second-generation Chevrolet Corvette. His collaborator here was the very talented Larry Shinoda, later designer of the Boss Mustang. Indeed, to Lapine the V8 928 was no less than a better-engineered iteration of the American muscle-car ideal. Since Bill Mitchell was apprentice to the ineffable Harley Earl, chief wizard in GM's den of kitsch, it is not too far-fetched to see a surreal connection here between fifties fins and eighties Porsche. One day I will write a speculative semiological essay connecting the whale-tail 930 Turbo to the 1957 Oldsmobile Holiday. I mean to say, things are rarely as they seem.

Lapine, who dominated Porsche design (with a lower-case 'd') until his retirement in 1988, was educated in what he called the University of Hard Knocks, although I think the novelist Nathanael West had said this first. He was an affable, pugnacious and likeable man who defined the Porsche aesthetic as having 'the winning looks that weapons have' although I do not know whether this was a conscious reference to Dr Porsche's earlier assistance with Hitler-era flying bombs.

But Porsche was beginning to acquire the desirable look that luxury goods possess. At some point in the eighties, it leapt the species barrier and became more than a car. Egregious period figures who could not translate *Luftgekühlt* began to drive them.

Examples? Wally Olins, the man who made 'corporate identity' a part of business lore, had a most distinctive 911 in vicious lime green. It was his calling card. Roger Seelig, the Morgan Grenfell banker who was fingered in the period-defining Guinness trial of 1987, brought his new 928 to my house. I told him not to leave it outside as it would be vandalised. But with a Master of the Universe's insouciant arrogance, he did, and it was. Of course, this 928 was painted Guards Red, the same colour as the banker's braces, an incentive to vandalism if ever there was.

Naturally, there are good commercial reasons for expanding the product line, as Volkswagen itself had added the epochal Golf to the epochal Beetle, thus sourcing the swaggering imperium that VW is today. But Porsche's range extension did not at first work. It is said that when one manager arrived at Weissach's *Entwicklungszentrum* in the late seventies, a chart on the wall showed a timeline with the 911 becoming extinct in 1981 and the 924 and 928 stretching into infinity. Of course, the front-engine coupés long predeceased the famous *Sportwagen mit Heckmotor*.

Maybe the Neunelfer will, like the Boeing B-52, survive to be one hundred. Maybe it will be the first industrial product to live forever. Of course, we should be grateful if it does, but that survival is due to philosophical compromises made in the eighties, that decision to diversify and stretch the meaning of 'Porsche'. It's a melancholy thought for purists, but the lumpen Cayenne is the most commercially successful Porsche ever.

Still, one day soon I will buy a 911. I mentioned this to that most perceptive of critics, my friend Paolo Tumminelli. 'Are you sure?' he said. 'In Germany they are only driven by grey-haired men with gritted teeth, or by women who drive their dog to the coiffeur at 50kph.' And then I had to remind him that I was a product of the eighties myself.

UNIVERSAL AND TIMELESS

No one ever thought the Porsche 911 'beautiful'. Yet it looks likely it will be a joy forever, as beautiful things are said to be. The end of production is not nearly in sight. On the contrary, the original design has shown itself capable of continuous evolution: proof of conceptual excellence.

The Jaguar E-type, however, has often been described as the most beautiful car ever made. Not least by Enzo Ferrari, never a designer himself, but an agitator of men and *padrone* of more mechanical beauty than anyone who has ever lived.

There is an E-type in the permanent collection of New York's Museum of Modern Art. Meanwhile, there are waiting lists for the latest Porsche. Compared with the 911, the Jaguar seems an antique, a museum-piece King's Road chariot of the sixties for men with flares and women without bras. Yet its haunting loveliness has hobbled every subsequent Jaguar designer, none of whom has been able even to approximate such ineffable beauty. I don't think you will find waiting lists for the F-type.

'Zeitgeist' is the term philosophers use to describe how contemporary products and events mystically share characteristics that bind them to their moment in history. But 'the spirit of the age' is not a useful device for explaining these different cars. They leave any analytical methodology floundering because each was the result, not of a dedicated research programme, but of a sequence of accidents and opportunities. And market research played no part. Instead, it was what the poet called 'the madness of art'.

Graf Albrecht von Goertz, a slightly dodgy 'count', who had worked for the genius-charlatan-hustler Raymond Loewy, in the US, and on the side designed the BMW 507, is an important source of the 911. On his return to Europe, he worked as a consultant for Porsche on a replacement for the 356. His proposals were rejected as being too Goertz-bling and not enough Porsche matter-of-fact.

But Goertz later persuaded young Ferdinand Alexander Porsche to leave Ulm's Hochschule für Gestaltung, sacred source of 'systematic' design, and join the family business. The shape of the 911 is Ferdinand Alexander's work. Clearly, he acknowledged history because its lines reflect Erwin Komenda's original Volkswagen profile.

But it is a more athletic, racing beetle, even if photographs of the 1963 original sitting high on its thin tyres give more an impression of a family limousine than a racing car. Indeed, the original brief insisted the 911 should have space for a bag of golf clubs.

Artistically, the essence of Ferdinand Alexander's design is geometry, a discipline revered at the Ulm school which also produced Dieter Rams, whose Braun electrical products later so inspired Jony Ive at Apple. Start doodling with overlapping ellipses for roof and hip lines and with semicircles for the wheel arches and you will soon have your own drawing of a 911's profile.

There is a BMW connection for the Jaguar as well. The lines of the XK120, the car that established the firm's reputation for dangerous proto-MeToo! suavity, were borrowed from the prewar BMW 328. And here we must ask where, in the case of the heroically opportunistic William Lyons, inspiration ends and plagiarism begins.

But when the aerodynamicist Malcolm Sayer began to exert his influence in Jaguar, other factors became involved. And they were not all sourced in the wind tunnel. Sayer admired the 1952 Alfa Romeo Disco Volante by Carrozzeria Touring of Milan. This is another astonishing composition of segments and sections of circles which indubitably influenced the E-type.

The Porsche and Jaguar engines are, in aesthetic terms, both deeply revealing of contrasting national preoccupations. In the German car, the engine is as anonymous and as unobtrusively functional as a fridge motor, but the Jaguar's is theatre and the stage curtains are that enormous, sculpted bonnet. Opening it is a process of concealment and display and if that is sexually suggestive, then so much about the E-type's aesthetic is.

It was an engine designed to be admired: look upon those three SU carburettors, the polished cam boxes and the very self-conscious triangular air filter and don't despair, but thrill. The exhausts, like the power bulge, were artistically emphasised to create drama.

In plan form the Jaguar, so often described as 'phallic', is a flat rectangle. And it is not so much a very long bonnet as a very short cabin, an arrangement which creates startling proportions. From front three-quarters, it's evident that those voluptuous wings are nearly pure cylinders. Ian Callum once told me, 'Malcolm Sayer designed by geometry … it is not free expression'. In Callum's analysis, the trailing edge of the E-type bonnet is a pure radius with a known mathematical value.

The Jaguar has better details than the Porsche. There is that power bulge, a sculptural device which adds an irrational complexity to the geometry of the bonnet. The chrome bar splitting the perfectly proportioned air-intake has nothing to do with aerodynamics and everything to do with a stylist's genius. Both Porsche and Jaguar have wrap-around rear bumpers, but the Jaguar's are somehow more lascivious. Callum said, 'It's all about putting just enough style into a car to make it fascinating'.

As we look at these old cars, it's impossible not to reflect on how little is ever truly new in matters of design. Some early 911s had engines recognisably related to Hitler's wheezing and puffing Volkswagen. And those distinctive horizontal rear lights on the E-type? They are standard units borrowed from the Jaguar Mark X, where they were deployed vertically.

Despite their differences, the Porsche and the Jaguar have much in common, aesthetically speaking. In each, the details are in harmony with the whole: you can instantly recognise a 911 or an E-type from a fragment alone. Rather like an organism, the entirety of each car evolves from its elementary parts. 'Classic' means the best of its kind, which is why major golf and tennis tournaments are called classics. It takes time to become a classic. Anyone talking about an 'instant classic' has inhaled too much PR. It took the 911 and E-type over half a century.

I do not think the Porsche and Jaguar designers were aiming at timelessness, even if that is what each achieved. But time is a cruel mistress: the beautiful E-type is dead while the more matter-of-fact Porsche lives on.

A GENUINELY MODERN PLEASURE

I once met the test pilot who was the first man to use an ejector seat. I mean the very first. Despite or because of the harrowing experience of voluntarily detonating an explosive charge beneath his bottom during high-speed flight in a jet aircraft, he was remarkably phlegmatic. Perhaps he was deaf.

The first man to reach 200mph on land must have had a similar fearless equanimity. This was Henry Segrave, an Old Etonian. Someone less fussy about cliché would have called him 'dashing'. Anyway, it's a truism that Old Etonians with their polished, but very sharp, elbows always get there first. Speed records included. And the number 'two hundred' had some mystical association for him: in 1921 Segrave won the Junior Car Club's 200-mile race. Two years later in France, he was the first British driver to win a Grand Prix in a British car, a Sunbeam from Wolverhampton.

I often wonder what Segrave's 200mph experience must have been like. Physicists recognise four types of motion: oscillation, rotary, reciprocating and linear. Pistons, shafts, planetary gears, thrashing chain-drive, primitive tyres getting very hot indeed. On that March day in 1927, Segrave endured them all for long enough to achieve an average of 203.792mph over two runs in opposite directions at Daytona Beach, in Florida.

Four years later, Aldous – *Brave New World* – Huxley explained that speed was 'the one genuinely modern pleasure'. After all, flight had been a regular, if minority, experience since the Montgolfier brothers' original hot-air balloon ascent, in 1783. As if to confirm that speed is addictive, Huxley added it is 'a new drug'. Amphetamines were synthesised in 1887. Of course, they are known colloquially as 'speed'.

What drug Segrave was on at Daytona we cannot say, and no one thought to ask him before he was killed in the Lake District in pursuit of the water speed record, three years later. But strong tranquillisers would surely be prescribed to anyone contemplating a ride in his car.

With nice insouciance, an OE characteristic as predictable as sharp elbows, it was called *Mystery*. Power came from two Sunbeam Matabele V12s, first (briefly) employed in an accident-prone French airliner of 1920. In Segrave's car, each 22.5-litre unit developed 435bhp. And they were mounted fore and aft, so, at least, equitable weight distribution was not a problem.

Starting was, apparently, a bit of a fuss. The rear engine was brought to life by compressed air. This was then, if you were lucky, coupled to the front engine via a friction clutch. Once both engines were running, they were synchronised by a dog-clutch, so no slippage might occur – slippage being a problem when you are messing with nearly a thousand bad-tempered horses. And off you went, nicely synchronised, to a double ton.

No one living can remember the original noise, but it is fair to speculate that it must have reached damaging levels, which can lead to personality change and violent reactions. I dare say the high and low frequency vibrations were an excitatory stimulus. Brain scientists know that the result can be a shift in the perception of horizontals. This means that as Segrave wrestled with the hot, noisy contraption that was *Mystery*, he was quite literally not seeing straight. The Florida horizon was wobbling at 200mph. On one run, Segrave drove *Mystery* into the sea to achieve a desired retardation, which the brakes had failed to provide.

More sedate speed experiences of 1927 included the *Royal Scot* steaming travellers on the London Midland and Scottish Railway to a genteel 110mph. More extreme included Mario de Bernardi's 297.7mph air-speed record in Venice, flying a Macchi M.52 seaplane built for the Schneider Trophy races. A contemporary Rolls-Royce Twenty could manage, perhaps, 60mph.

Now, 200mph seems almost commonplace. The Ferrari F40 of 1987 was the first production car to claim it, but since only 1,315 were ever built, this is 'production' only in the sense that cut-throat Sardinian bandits may be described as modern Italian businessmen. But I have seen 300km/h (or 187.5mph) in an Audi limousine. And I am not a brave person.

Henry Segrave's 200mph was a stirring mixture of personal bravery and technical audacity. The thrashing chains were protected by shrouds to contain the sort of accident that now and again led to drivers being decapitated. I don't imagine Segrave thought much about this. Nor about crash protection or fire prevention. And then the circus moved on. In 1935, speed records began at Bonneville, in Utah, where the surface of the salt lake was more nearly scientific than South Atlantic Avenue in Daytona.

But I also suspect Segrave did not think about the aesthetics of what he was doing. Never mind his vibrating eyeballs making disturbing patterns of the horizon, the valorisation of speed was a matter of culture as much as technology. Speed alters our sense of space. And the car is speed's most willing facilitator. The car democratised modernity: in the twentieth century, motion became more important than territory or domesticity. Of course, all the signs are that in our anxious and insular moment, territory and domesticity are being re-prioritised. But that just makes Henry Segrave's 200mph folly all the more touching. Nowadays, heroics are best understood in terms of loss.

VULGARITY JAG

One of the characteristics of our historical moment is inflation. For a long time, everything seemed to be getting smaller. Now everything is getting bigger. In 1959, absolutely the smartest appliance you could own was a miniature Sony television with an 8-inch screen. Now you are not even in the game unless you are staring in contented bemusement at an idiot's lantern 6 feet across.

It is the same with job titles. Once there were 'commercial artists'. There were Abram Games and Saul Bass and Cedric Morris. The last, an interesting fellow, the Ninth Baronet, who set up the East Anglian School of Painting and Drawing and taught Lucien Freud. Morris was not too proud to do magnificent posters for BP. Then professional inflation forced commercial artists to evolve into 'graphic designers' who, in turn, ceded to 'branding consultants'. Did this mean anything different or better? Was it signal or noise?

So, when I read that the excellent Marek Reichman of Aston Martin had declared 'we are not car stylists, we are design engineers', I paused for thought. Reichman is one of the very best of the current bunch, although that may be a little like saying he is the finest lyric poet on the staff of *Yellow Pages*. Be that as it may, I don't see what is wrong with 'styling'. Why be ashamed of giving emotional expression to dumb materials? Styling is the dress of thought. And in these brutal days, we need more of it.

God laughs at our plans and my guess is He absolutely roars with thigh-slapping howling hysteria when shown leaked motor industry forecasts. But despite the imminent changes in propulsion technologies and legislative or fiscal environments, the car is going to be with us in a recognisable form for about as along as anyone dares imagine. And that form must be invested with huge steaming dollops of style if it is to have any meaning to users. Or, I should add, acquire anything like the cultural status of a '57 Thunderbird.

Instead, when I look around at the motor industry, I just see all of them in a pantomime lock-step gavotte of futility, searching for meaning outside the day job. Instead of encouraging commercial artists to dream in good proportions, as artists do, they have sent for the branding consultants, when what they need are more aesthetes. Thus, Marek the Excellent has teamed up with Satan to do some Aston Martin brand extension. Satan never laughs at anybody's plans (he finds the present moment hilarious enough), but he has occasionally made the point that it is always best to stick to the knitting.

I don't want Gordon Ramsay to design my suit. To be truthful, I wouldn't want Gordon Ramsay to cook my dinner (unless he was wearing one of the Met's spit masks). But nor am I really certain that I want an Aston Martin branded apartment. I dare say there is a dwindling group of gas-entrepreneur kleptomaniac oligarchs, in Vladivostok, who would care for such a thing, but snuggling up to them is not in my view how culture advances. What next for Aston? A limited-edition colour-coded Bond-era Walther PPK in a Bottega Veneta holster in the glovebox of your DB11 along with the spare bulbs?

Unsurprisingly, Rolls-Royce, presently on a vulgarity jag that would shame Heliogabalus, has hit a very rich seam of kitsch in this mysterious voodoo of annexing meaning by making connections beyond the motor trade. Rolls-Royce customers have recently been invited to admire a Fabergé egg in purple enamel on a base of white gold. Within, the Spirit of Ecstasy is hand-sculpted in frosted rock crystal and revealed by a lever-operated mechanism which parts the halves of the egg rather in the same way that the topless Miss World used to burst out of a fibreglass swan bobbing on a hotel lake. I am only surprised that Rolls-Royce branded sickbags in faux unicorn hide are not included in this hideous 'bespoke personalisation' experience. The original Russian Imperial Fabergé eggs, by the way, were commissioned by the Romanov family. And just look what happened to them.

In a precarious world it is easy to understand how attachment to the past might be comforting. Indeed, Aston Martin has revived Superleggera, even if the cars no longer have any connection with Carrozzeria Touring's lightweight construction. But attachment to the future is surely more compelling. And I know Aston Martin has interesting plans here. (Sound of Divine laughter.)

Look at Maserati. People old enough to remember the 250F and Birdcage no longer have their own teeth and, wisely, Maserati has resisted any temptation towards a period costume drama to tempt the dentally deficient. But the reason Maserati has not been doing terribly well commercially is very simple. The cars are not beautiful enough. You could sadly say that of Alfa Romeo too. And, alas, the same is true of Jaguar, whose poor commercial performance can surely be attributed not to technical deficiencies, but to lacklustre 'styling'.

Beauty is, of course, the job of the stylist. Ferrari's Flavio Manzoni taught me the expression '*tentare non nuoce*': it doesn't hurt to try. When all the agonies of electrification, pollution, congestion and allocation of scarce resources have been managed, the car will remain a primarily aesthetic object. And I think it will need stylists as much as it needs design engineers or branding consultants.

GIRLS, PAINT AND SPEED

Most days I walk past the Ferrari showroom, on the dangerous frontier between Knightsbridge and Belgravia. A few hundred yards further on, McLaren has premises of its own on the ground floor of the bullying and overbearing Richard Rogers building at One Hyde Park, which will one day be fully understood as that once great architect's tragic fall from grace.

On show in each are cars that are very near the ultimate proposition of the automobile (at least as presently understood). So, I am fascinated that both Ferrari and McLaren actively encourage customers to tinker with their finely wrought products. McLaren's Special Operations will carry out bespoke orders and Ferrari's Atelier invites you to specify finishes and fittings that were not the first choice of Pininfarina. It's as if Heston Blumenthal, whose flagship restaurant is just opposite Ferrari, suggested you bring your own BBQ sauce.

For me, Ferraris are just fine exactly as they left the factory. Presidente Luca Cordero (let's agree to forget the di Montezemolo business which is just like saying 'of St Albans') is fond of quoting the Paolo Conte lyric 'A sports car must smell of girls, paint and speed'. Exactly so. But it seems nowadays a significant number of customers want their sports car to smell of other things as well.

The urge to customise is deeply inimical to anyone who has made a fetish of design. The whole point about design is that it is meant to be correct, inevitable, true and unimprovable. Of course, culture is dynamic, and no such state of affairs is ever possible, but you have to believe it might be. As Le Corbusier insisted, industrial production demands standards and standards lead to perfection. Or that's the theory. The customising urge suggests the theory does not work.

But while at one (rather purist) level I disapprove of customising, at another, I find it fascinating as an act of rebellion and a statement of self-identity. The psychology of customising is a curious one. On the one hand, it is entirely rational (possibly an urgent necessity) to seek improvements in a Moskvitch M-412. I once saw a robustly customised example in Havana: artisanal flared arches, an all-over gold paint-job, limo-tinted glass and big wheels. In the conditions of privation you find in Cuba, such an exercise was a powerful expression of enterprise and yearning. But why would you want to gussy-up a Ferrari 458 Italia which, so far as gussying goes, is pretty much as all-out gussified as a machine can be?

Like tattoos – I will not say 'mutilation' – which in several ways it resembles, customising has its origins in the demi-monde, specifically in the Californian counterculture, which emerged circa 1960 between the drug-crazed beatniks and the radically different drug-crazed hippies who followed. The presiding spirit was Ed 'Big Daddy' Roth, whose career began with T-shirts, but soon evolved into the Weirdo style that the magazine *Car Craft* celebrated in 1959. From his workshop at 4616 Slauson Avenue in Maywood (a Los Angeles neighbourhood with aircraft and automobile traditions), Roth built the 'nutty-looking, crazy-baroque' cars which Tom Wolfe brilliantly memorialised in his *Kandy-Kolored Tangerine-Flake Streamline Baby*.

This book, which I read on my knees in school maths lessons, when Pythagorean Identities did not seem to be quite enough diversion, was part of my aesthetic education. And I reverently studied its holy writ as I scrutinised US hot-rod magazines for visual support. I loved the dropped axles, the exposed Roots superchargers, the sculpted and chromed lateral exhausts, dog-dish hubcaps and sinfully lustrous candy-apple paint with blinding clear-coat lacquer. I liked the sheer commitment involved in removing a car's roof. And I loved the finishes. Hot-rodders might not have read Erving Goffmann's *The Presentation of Self in Everyday Life*, but they wrote the standard work on visual semantics.

Some years ago, I went to the Max Power show, in Birmingham. Absurd and horrible? Yes. Magnificent and moving? I think so. There is something infinitely touching about someone prepared to spend £30,000 on turning a drab Citroën Saxo into a cult object. Sure, Max Power stank of stale beer and fresh testosterone. Meanwhile, a personalised Ferrari and a customised McLaren are surrounded by the fizz of sparkling wine and a whiff of Prada scent. Otherwise, there they are with the Saxo.

Wherever it occurs, in Knightsbridge, Los Angeles or Birmingham, and whatever the result – a tricky Ferrari, a Frankenstein Ford or a Citroën with a glandular disorder – customising expresses an essential truth: cars are potent mediums of self-expression. That, ultimately, is the source of their wonder.

'MODERN' IS HISTORY

I blame air travel for a lot of things, including the present condition of car design. Travel used to be an exciting stimulus, a romantic privilege, but since the 747 democratised flight in 1970, it has become a predictable chore. The industry's designers travel a lot, but the sort of travelling they do is the sort that narrows rather than broadens their horizons.

The same plush, but heartless, international hotels, chosen because corporate hospitality has negotiated a cost plan. Airport lounges in which all references to locality or national identity have been extinguished in a dull blur of acceptable mediocrity. These are environments lacking the intellectual energy to make a wind-chime tinkle. This is why, nowadays, a Jaguar looks like an Audi and all small cars are identical.

I liked it when car designers worked in broader strokes on bigger canvases. And the biggest canvas of them all was the sky. Once, aviation was inspirational. Marvellous that the aerodynamicist Paul Jaray designed Zeppelins and seaplanes before he founded his Stromlinien Karosserie, in 1927, changing the industry's concepts of how an automobile should perform and look. Jaray also designed landing gear and, later, became an academic at Zurich's mighty Institute of Technology.

Or what about the piratical boldness of Alex Tremulis, who installed a 275bhp Air Cooled Motors 'Franklin' engine (which had seen service in the Bellanca Cruisair and Sikorsky S-52 helicopter) in his doomed 1948 Tucker? Later, Tremulis had his handwriting in the Boeing X-20 Dyna-Soar that predicted both the comedy *The Jetsons* and the NASA Space Shuttle. He worked briefly for Ford then, in 1978, through his own consultancy, produced the splendidly funky Subaru Brat pick-up. That's an eclectic portfolio.

The intellectual poverty and creative limitations of the current generation of car designers is a recurrent riff of mine. Every manufactured object reveals the beliefs and preoccupations of its maker. Or the lack of them. I wish I could show you the brochure of the 1953 Lincoln Capri I have before me, since it reveals an awful lot of beliefs and preoccupations. I can see a glorious convertible in Gordon's Gin green with blindingly white upholstery which must, surely, be plastic rather than leather.

One spread in the brochure is titled 'How Modern Living Sells', another 'Here's what Active Moderns Buy'. Active Moderns, it appears, buy A-line dresses (if female), live in courtyard houses with ragstone feature walls, take skiing holidays, have fitted kitchens, and own a streamlined fridge. There was a happy time when such things were signifiers of civilisation.

But not only did the '53 Lincoln connect with the cultures of architecture and design, it also won the top four spots in the production car category of the 1952 and 1953 Carrera Panamericana. In any case, the name Capri reveals, for a car conceived in Michigan, a plangent yearning for exotic sophistication. Yet another spread is called 'A New Insight into Mobile Interiors'. Featuring the motif of coloured acrylic balls, it could have been taken from a Herman Miller catalogue selling the Charles Eames Hang-It-All coat rack, mid-century modern's very best memorial.

But 'modern' is now history. Mid-century was the last moment when material culture was composed of solid objects that might inspire car designers to spasms of imitation (Harley Earl stealing the tail fins of a Lockheed P-38 for his Cadillac, or a rocket nozzle for a Buick grille) or inspiration (Carlo Mollino making his Bisiluro Le Mans racer an aeroplane that did not leave the ground). The great curiosity of our contemporary moment is that the most exciting technologies are hidden.

Even in aerospace, the big technical advances of today are invisible. To a large extent, the practical problems of aerodynamic form and lightweight structures are well understood. The value of a Boeing 787 is half hardware and half navigation-guidance-control systems. And fifty per cent of the latter is software. True, the people who write code for Rockwell Collins have to understand the hardware, but an essential link between the soul of a machine and its body has been lost.

It's all about behaviour, not appearances. The next aerospace technology will be craft-to-craft comms. Boeing is working on swarming systems for drones. There are clear applications here for ground transport: your motorway queue will soon be informed by an intelligence higher than the man in the racing Transit. Or take the Lockheed Martin F-35 with its 'distributed aperture systems' (DAS: which I think means 'a lot of cameras') for enhanced situational awareness.

The DAS system sees through the aircraft's structure, feeding data into the pilot's virtual reality helmet. His craft becomes intangible. Just think what that might mean for driving. Meanwhile, my man at BMW tells me they are very excited in Munich about getting GPS data down to centimetre accuracy which, when achieved, will make autonomy a safe possibility.

Designing a car to penetrate the air at 100mph brought beautiful results. As I imagine our designers discussing swarming systems in the first-class lounge at Shanghai Pudong International Airport, I find myself more than ever certain that the best of automobile design is in the past. A line you are never going to hear? 'Would you like to come for a swarm in my car?'

DELOREAN

Living and working in central London, whose busy vices would bring Sodom into disrepute, raises all sorts of thresholds. Excitement, tolerance of crowds, anxiety, expense and fascination, for example. Concerning the latter, I can walk trance-like past a Ferrari showroom full of shiny treasures without looking up from my smartphone. Although I confess a totally matt black Rolls-Royce Phantom with backlit cursive numberplates did briefly divert my fragile attention the other day. Whatever, the game is continuously being raised.

Then one night, strolling home through a quiet Knightsbridge mews whose landmarks are a bespoke scented-candle boutique, a cute pub with candles of its own and a frock shop in an old dairy whose standout feature is its expensive anonymity, I saw something truly distinctive. It was dark and wet; the distant silhouette was not immediately recognisable, but the stance was impressive and the effect unique.

Curiously, there were no other cars parked nearby, so, as I approached, there was an opportunity to consider the options without distraction. I'm a good car spotter, but I was uncertain. Getting closer, I realised it was a DeLorean. I stood and admired, even if its aesthetic faults are glaring. It sits too high and the wheels are wrong, the details are seventies, but not in a good way. Yet the Giugiaro origami shape (an original *disegno* rejected by Porsche) is astonishing and the stainless-steel surfaces a delight and a temptation. I wanted to touch, but quickly noticed the Russian plates and thought this curiosity might be rewarded by a rapid and ruthless response from protective Spetsnaz retirees on retainer to a local oligarch with a curious taste in cars.

There is a lot of curiosity in a DeLorean. Stainless steel is anything but stainless and, in fact, attracts blemishes as surely as un-dyed, river-washed linen. The manufacturers advised against using a metal scouring pad to clean it since this might leave minuscule ferric traces prone to rust, which eventual effect sounds rather cool to me. Those rear window louvres were one of the most complex mouldings of the day but made a retrospective view of low-speed manoeuvring impossible. At least, impossible without expensive incident.

The gull-wing doors were almost insupportably heavy and did not allow for proper opening windows, so owners in hot climates learnt it was acceptable to drive at low speeds with them open. (I am told that the same remedy was often employed in California by suffocating drivers of Mercedes-Benz 300 SLs.) At a late stage in its development, the DeLorean's engine was swapped from a mid to a rear position, conferring on the car a dynamic that all users of pendulums will know and fear. Additionally, it was slow and heavy with spectacularly awful ergonomics: the centre console was so high that reaching the gear lever was like fiddling above the lintel of a door to find a hidden key.

But these are small-minded niggles because, in so many ways, the DeLorean is a thing of wonder. Even if it is poisoned wonder. Was there ever a car so full of promise, but which achieved such calamity? A car, like any product, reflects the character of its creator. It is one of my definitions of what 'design' means.

So, like the car he made, John Zachary DeLorean had a lot of character, most of it bad. To improve his non-verbal communication, he had cosmetic surgery on his jaw the better to conform to a stereotype of thrusting executive manhood. He became adept, according to court proceedings, at forging the signature of his collaborator, Colin Chapman. The *Guardian* called him a 'con-man' (although not while he was alive and energetically litigious).

In the year the DeLorean Motor Company went bust (taking about £100m of public money with it), he converted to fundamentalist Christianity during a kitsch baptism in his spiffy Californian pool. Later that same year, he was bust for a $24m cocaine deal, escaping prison by claiming entrapment by the FBI. He sued everyone promiscuously. He wore flares. His last business venture was a scam taking deposits for new DeLorean branded watches that did not exist. This from the man whose plan it was to make an 'ethical' sports car.

You could say: if these are ethics, we can do without them. But there is something compelling about such awfulness. Not every great artist had impeccable manners or morals or taste. Caravaggio and Gesualdo were murderers, but we do not admire their paintings or music any the less. Henry Ford was a vicious bigot and anti-Semite. Ferdinand Porsche mentored Hitler. Buckminster Fuller's futuristic Dymaxion car killed its test-driver. André Citroën lost his fortune in casinos.

Yes, you can read cars and interrogate their character, but there is no moral equivalence between being interesting and being virtuous. Bad people can create great art. That DeLorean I saw in the Knightsbridge mews looked rather wonderful. It's a car that tells a story ... and stories always fascinate.

DEATH SPIRAL OF THE SPORTS CAR

As a boy, my coordinates of desire were, at least until I discovered girls, described by points on the map of car culture. I wrote earnest letters to sales and marketing departments everywhere and got fabulous, glossy brochures in response. These I dutifully indexed and filed, building up a mental library of seductive imagery still with me today.

My favourites came from GM. The Pontiac division eschewed banal photography (perhaps because the camera has this thing for the truth) and instead employed an illustrator called Art Fitzpatrick, whose insanely glorious gouache renderings gave an already low and wide Bonneville the proportions of a tennis court. And the cars were placed in settings that were, to a suburban provincial youth, impossibly glamorous: beaches, country clubs, swaying palms, white clapboard colonial-style houses.

Fitzpatrick's passengers were handsome couples – he with a cravat, she with a headscarf – always seemingly poised on the abyss of crazily tumultuous sex on the pleated Naugahyde wipe-down front bench-seat.

While Art was stirring his pots of paint and using bizarre proportions and expensive Kolinsky sable brushes to extend our range of imaginative possibilities, the future novelist Elmore – 'Dutch' – Leonard was writing advertising copy for trucks in the GM Building on Detroit's West Grand Boulevard. Oh, pleasant exercise of hope and joy. Bliss was it in that dawn to be alive.

But does anyone produce brochures any more? Are genius crime novelists writing Silverado HD copy? An earnest fellow in the retail motor-trade recently told me, not without enthusiasm, even as I was yawning within, that nowadays 'pre-purchase data-gathering is all online'. In any case, the young are not interested. Smoking gun? A smoking exhaust pipe is now damning evidence of social irresponsibility.

Car culture is unquestionably coming to its end. I have always thought of car design as essentially similar to art: a professional activity with its disciplines, techniques, traditions, studios, maestros and apprentices. And all bent on providing aesthetic rewards for customers. But that version of art is over as well.

For example, it took me a while to notice, but the 'sports car' is in a death spiral, although that might be too positive a term. In the US, the Ford Mustang has not excited uncontrollable national outbreaks of hysteria. In fact, they are hard to sell. Meanwhile, the blobby Macan is easily Porsche's most popular product. Of course, we have all watched the gradual popular acceptance of the

SUV, but without realising that this format is now the norm and everything else, especially an open two-seater, is an aberration.

And the entire German car industry is in an existential crisis which makes the death struggle of Laocoön look like a peck on the cheek from a flirty ex. Those fine traditions of lengthy apprenticeships and secure employment which cultivated the great technical skills that gave German cars their admirable aura of quality suddenly seem very, very expensive. And irrelevant.

Do you continue to invest in old-fashioned expertise to develop internal combustion engines, but starve EV research? Or do you commit to electricity and undermine the precious essence of your hard-won brand values, making redundant all the wonderful stuff you have spent the last century promoting? Audi and Daimler say, 'Don't know', but have laid off tens of thousands of skilled workers in any case. Tell me if you know of anyone planning a new V12 (apart from Gordon Murray).

And while a car is more than the sum of its parts, in a mystical sense the dense assembly of beautifully conceived and finely executed mechanical components created a settled gravity which, experience tells us, people enjoy. With a car, aesthetics are as much felt as seen. It's a symphony of disparate elements. But electric vehicles have far fewer components than internal-combustion-engine cars. Electric vehicles are like fridges. Meanwhile, Mahle makes pistons. Would you say, in stock-market terms, if Mahle is a buy, hold or sell?

And, incidentally, forgetting pistons for a moment, most of the batteries used in electric vehicles are supplied by Asian companies. I believe that great cars of the past have almost always been inspired by a designer's response to a power source based on cams and gears and cranks and valves and chains, a living thing of smells, heat and noise. I don't see anyone getting incontinently excited by a charmless 23GWh cell supplied by Panasonic, Contemporary Amperex Technology, BYD, OptimumNano Energy or LG Chem.

The problem here is that complaining too much about evolution makes you look pre-Darwinian and, as a (bruised) modernist progressive, that's not a look I care for. Nor do I own a gorilla suit and drag my knuckles. And it's chastening that the only pro-internal-combustion-engine campaign I know of is from the intellectually inelegant Alternative für Deutschland party, whose pro-diesel rants might as well be pro-coal or pro-slavery.

I am afraid the old car culture really is at death's door. No one is going to write 'We'll have fun, fun, fun 'til daddy takes the Tesla away'. Indeed, no one writes that sort of song any more. But this simply means the treasures car culture once produced now become more precious.

THE END OF DESIGN

So far as I am concerned, the last great innovation on wheels, excluding the Daimler-Swatch collaborative's smart (which I adore), was Robert Plath's 1991 Rollaboard. Plath was a Boeing 747 pilot, with Northwest Airlines, who perfected the wheeled suitcase. Just wide enough to fit down an aircraft's aisle, he got stability and manoeuvrability issues sorted out by having only a single pair of wheels at the back. Plus, that ingenious rigid handle that extends and retracts as crisply as landing gear.

Of course, I detest wheeled suitcases on the basis that (a) morally, if you can't carry it, don't bring it; and (b) practically, people who use them spontaneously lose all spatial sense, civic decorum and become glazed-eyed and moronic, and routinely run over your Gucci loafers. Still, what amazing ingenuity. As someone pointed out, isn't it odd that we had a man on the moon and aerosol cheese before anyone thought of putting a little bogie on a bag?

I am no longer certain that creativity is the driver in car design. And nor is Thomas Ingenlath, formerly of Volvo. Following well-publicised neuro-aesthetic experiments with dry electrodes (things could get dull in Gothenburg), Ingenlath concluded that, in the premium sector, design was no longer so important in establishing a competitive advantage. This is a bit disingenuous, since Ingenlath spent several years more-or-less successfully designing and dragging the artistic shambles that was Volvo into a credible simulacrum of a premium product with nice gloss and sculptural presence. But, essentially, he may be correct.

No one is anticipating any radical design innovations in the future, aesthetically speaking. On the contrary, the consumer is stupefied to a condition of catatonic ennui by parades of boring 'concept' cars pimped at motor shows. Once, such things offered genuine excitement and vision. When in the fifties General Motors moved its *Motorama* roadshows around the United States, their arrival in the provinces was, quite literally, sensational. You are in a dreary US city in the flyover states where the biggest concession to cultural interest is the orange roof of a Howard Johnson motel and someone shows you a Buick LeSabre styled to look like an interstellar transit pod with hot jets and penetrating probosces. That is something which extends your cognitive horizons further than *Reader's Digest*. And next year, it would be different. But today, there is a settled opinion of what cars, especially cars in the premium sector, should look like. Will Audi ever change much? I doubt it.

This train of thought prompted another: what were history's most innovative premium car shapes? Which designers and which machines established a new way of seeing? Who built automobile architectures that created templates which successors had to imitate or be stigmatised as retardataire? In the belief that there is nothing quite like a list to start an argument, here is mine:

1. Harley Earl's Buick Y-Job. This 1938 project was the original 'concept' car, anticipating the visual language of the fifties: smooth, integrated, sculptural and modern. This was design as a synoptic idea, not a dress disguising a contraption.
2. George Walker's '49 Ford. The first authentic postwar car. A wholly conceptualised consumer product, the epochal '49 Ford has more in common with a streamlined fridge than, say, a contemporary Maserati 4CLT.
3. Pininfarina's 1955 Lancia Florida. This exquisitely proportioned concept established aesthetic rules for the European saloon that were not much tested until the aero fads of the eighties.
4. David Bache's 1970 Range Rover. Pure genius in that Bache did not just hit a target, he saw a target that no one else could detect.
5. Peter Schreyer's 1997 Audi A6. A disciplined evolution of Hartmut Warkuss's astonishing 1982 Audi 100, with a little avant-garde assistance from J Mays' even more astonishing 1991 Audi Avus concept, the A6 formalised Audi's language of bold understatement still spoken today.

And since then I can think of no car that has been a significant aesthetic step-change. Our millennium's design is still working on the aesthetic assumptions, smarty excepted, that were established the century before.

One reason for this, as Ingenlath says, is the increasing electronic content of cars. The thinking goes that, with the artistic issues settled (the latest Volvo XC90 is, sculpturally, no more than a cautious refinement of its predecessor), the creative challenge is about intelligent human conversations with gigabytes, not lascivious human desire for a voluptuous hipline. Well, maybe.

In 1992, Francis Fukayama wrote a book called *The End of History*, arguing that Western liberal democracies were so successful that no more discussion of politics was ever going to be necessary. Maybe we have now reached The End of Design since everyone seems to know what they are doing with no more debate required. Then again, maybe not. Soon after Fukuyama's predictions, global geopolitics exploded in entirely unpredictable ways and his book looked about as prophetic as *The Wind in the Willows*.

I think the world is ready for aesthetic surprises: the car industry is waiting for its Robert Plath.

EXHAUST PIPES

We have three hundred million neural pattern recognisers. These allow us restlessly to make fine judgements about the look of things. But aesthetics cannot be reduced to a cerebral wiring diagram or formulaic brain chemistry. To do this would 'drain enchantment' from the world.

Anyway, there is no need for concern: even if science claims a spurious sort of accuracy in its assessment of things, cultural factors intervene and muddle the perceptual equation with the uncountable influences of desire, charm and taste. When you look around, there's an awful lot of enchantment left.

Perhaps curiously, that highfalutin intro was inspired by a Bentley GTC's exhaust pipes. I had been driving this handsome and powerful car as hard as my limited skills and narrow Chilterns lanes allowed on a dark and dirty night. We parked at our destination and left the car for dinner. Returning to the Bentley in the morning, my neural pattern recognisers were surprised to see that of the four ellipsoid exhaust pipes, half were fake. Arranged in two pairs left and right, only one of each had soot. The other was a trinket.

I immediately thought several things. One was that, nearly ninety years after the foundation of the Bauhaus, an art school inspired by the 'logic' of machinery, the fiction of form and function has at last been exposed. Form never did follow function. Form just does exactly what its designer wants it to do.

I also wondered why exactly Bentley had done such a thing. Would GTC customers feel short-changed if their car had a mere two exhaust pipes? What was the precise reasoning that argued this case? Is there a checklist of necessary symbolism which circulates confidentially between designers? And is this same customer truly gratified when his appetite for expressive gestures is teased by a falsehood?

And I began to think how fascinating exhaust pipes are. Sigmund Freud had a wonderful expression: 'the narcissism of small differences'. Or, the more trivial the distinction, the more meaning we see in it. In the simple matter of providing a passage for burnt gas to be expelled from an engine, the automobile industry has created an entire abecedarium of different meanings. Some are delusional, others matter of fact, but my point is: you can learn a lot about the psychology of a car designer and his customer by looking at exhaust pipes alone.

I remembered the early sixties, when the first generation of Minis was causing delight. People might accessorise them with a shiny wood-rim steering wheel or, pertinent to this column, a little silencer with twin wide-bore tailpipes.

Of course, the back pressure from these oversized orifices was enough to threaten the puny 848cc A-series engine into a stall, thus their presence simply advertised the technological illiteracy of the owner. Still, he felt empowered and that's the important thing.

Meanwhile, do you think this crude after-market add-on was evidence of the fraudulent cynicism that contaminated the British motor industry in its twilight agonies? Maybe, but have a look at a new BMW X5 next time you are passing. Those beautifully finished chrome tailpipes with that nice lipped return around the circumference are add-ons too. They might have been designed in laboratory conditions in Munich, but their fundamental appeal is addressed to the same neural pattern recognisers that made Polynesian islanders attracted to colonists' beads and mirrors.

So, here is a question I never expected to ask. What are the greatest ever exhaust pipes? There are several contestants. The 'Little Deuce Coupe' that inspired the Beach Boys had, besides its competition clutch with a four-on-the-floor, a device known to hot-rodders as 'lake pipes'. These are unsilenced tubes for the expulsion of noxious car excrement, typically finishing with a gaping, almost sexual, flourish just aft of the front wheel and beneath the running boards.

Then there is the Indianapolis-winning Lotus 38 with two absurdly provocative horizontal tapered pipes like barrels of an Imperial dreadnought's gun turret running in reverse at 200mph. But the ultimate is surely Forghieri's spaghettini. Mauro Forghieri's 1967 Ferrari 312 was not a specially effective Grand Prix car, perhaps because the majority of the creative effort was invested in the conception and fabrication of gloriously complex exhaust pipes that express the beautiful violence of internal combustion with a sculptural power which, late at night, I might compare to Michelangelo. Deliciously, they were painted matt white.

Many years ago, I was involved in a corporate identity project for Volvo. We had a discussion about exhaust pipes, and all thought it a very clever idea that Volvo should express its mythic ruggedness by wrenching a 240's drinking straw of a tailpipe from out of obscurity behind the valance and make a bold, functionalist gesture of it. Of course, this idea was ahead of its time and came to nothing. But today, no one ignores exhaust pipes. In the designers' endless quest for more stuff to work on, what goes out the back has come to the fore.

GERMANS

What has gone wrong with the Germans? True, they have often suffered a reputation for coarseness. The Emperor Charles V, discussing levels of national sophistication not irrelevant today, said: 'I speak French to my mistress, Italian to my courtier and German to my horse'. During the Great War, the French called the Kaiser '*le chef des barbares*'.

But the Germans have – or had – a superb record in design. Nineteenth-century pedagogues developed the idea of *Hauptformen*. It's untranslatable, but means approximately 'significant form', that's to say there is an ideal shape for any product. This belief fed directly into art education and Bauhaus doctrine. Indeed, the Bauhaus had as its logo a sphere, a cube and a cone. In primary colours.

German domestic architecture in the first half of the nineteenth century was in a style known as Biedermeier, a name derived from the pseudonymous author of poems mocking the *taedium vitae* of the middle classes. Yet Biedermeier buildings are solid, finely proportioned and possessed of a certain timelessness. It's not too fanciful to see a direct connection to the crisp and airy 1961 BMW 'Neue Klasse' saloons of Wilhelm Hofmeister which made the company's fortune by exploiting these exact same qualities.

But look at BMW today and what you see is not clarity and commodity, but undisciplined, uncoordinated, directionless mess. Its product line looks like the aftermath of an explosion in a car park when someone has tried to re-assemble all the bits. Never mind the atrocity that is the Rolls-Royce Cullinan whose otherworldly grossness, we must assume, was given an enthusiastic nod by senior management who saw in it the ultimate expression of their degenerating corporate values, the core BMW range is, aesthetically speaking, no more attractive than a putrescent rat on a trembling stick. I recall thinking on first sight of the last generation 3 Series, dear God, it looks Korean. Today? If only it looked more Korean. When Kia makes more attractive cars than BMW, *Alles ist nicht in Ordnung*.

I am not alone in these solemn reflections. The entire German industry's fall from artistic grace excited the country's senior automotive journalist, Georg Kacher, into a splendidly fuming and intemperate piece in the normally sober *Süddeutsche Zeitung*. The Audi A8 with its idiotic gaping grille looks like a Bush-era Lincoln. The smart is Playmobil. The Mercedes-Benz B series is sexed-up ('*aufgepepptes*') Swabian Biedermeier. The BMW 2 Series is contemptibly simple, the 7 Series 'fails to be regal', but Kacher reserves his absolute raging and foaming contempt for the BMW X7 which is '*grobschlächtig*'. Untranslatable again, but emphatically not flattering.

Most great design is achieved with economy. The disciplines of restraint stimulate genius. However, today's markets are not interested in dignified economy, but in obscene indulgence. And the Germans have a cultural problem with luxury. I noticed this long ago with Wolfgang Reitzle's feet. The man who had overseen the beautifully well-mannered fourth-generation 5 Series was wearing those woven leather loafers you are most likely to see in Abuja brothels. Amazing. Luxury seems to drive Germans demented.

The design discipline that once informed BMW has been dissipated into meaningless model variations heading who knows where. The famous reticence of Audi is compromised by look-at-me flashes of vulgarian bling. Mercedes retains traditional gravitas, but its absolute, patrician design authority is being diluted by weird product planning.

It may be that the admirable historic German inclination towards systematic thinking cannot accommodate the excesses demanded by contemporary markets. The Bauhaus did not do ruched leather, but nor – then – did the Bauhaus need to sell cars in Shenzhen.

Impossible, really, to imagine an X7 or a Maybach being a product of the same culture that begat the Hochschule für Gestaltung which, in turn, begat Dieter Rams and his 'good design is as little design as possible'. Luxury stains whatever it touches. And when Germans opt for luxury, they get stained, tip towards excess, and that soon slides into irrecoverable decadence. Ask anyone who visited the Moka Efti Club on Berlin's Leipzigerstrasse. Ask Christopher Isherwood or Sally Bowles. Or look at an X7.

So, the answer to my introductory question is that the Germans have lost, or neglected, their very own principles. They have become design apostates. Instead of making rules, they are chasing markets. And chasing is rarely dignified.

I give you as evidence, the late Audi A2 (d. 2005), the last great German car. It was technically adventurous, uncompromised in execution, completely original, but also very evidently a part of an established design tradition. And it looked so very German: it had those wilfully odd proportions and peculiar general arrangement that characterised Richard Vogt's Blohm & Voss BV 141 reconnaissance plane or a Hanomag-Henschel F-65 truck.

And it was marvellous. The problem was the A2 nobly overestimated the public's taste. But the reality of the car trade is that no one goes bust underestimating it.

NOISE

We were having lunch in a favourite Soho dive with baffling noise levels, but huge character. I mean to say: Canaletto once lived in the building! His ghost is still there giving hints and tips on Venetian cooking. My companion was an old friend who knows a great deal more about driving than do I.

As he studiously navigated a path through a glistening pile of *fritto misto*, he explained that he had just been driving a Tesla, which was pronounced to be very impressive. Handsome, fast and comfortable, he said. So, I asked: 'Would you buy one?'. Hesitating for a moment while he theatrically removed an intractable bit of deep-fried calamari from his teeth, he scrunched up his brow and said, 'No'.

The reason, it turned out, was much the same as the frequent failure of many of the old design 'clinics' where customers were invited to an unventilated backwater of a hotel on the periphery of Heathrow, shown a de-badged car and asked questions such as, 'Does it look expensive?' Or, 'Does it look German?' No one ever asked the most important question of all. 'Would you dream about it at night?' In this way the 1985 Ford Granada slipped into the world with no one's approval.

The problem with the Tesla is the noise, or rather the lack of it. Nikola Tesla's rotating AC induction motor is a thing of wonder, making a magnificent job of converting electrical energy into propulsive mechanical energy. But electric motors are boring. They are all much the same. Exploding complex hydrocarbons is much more interesting. In a more innocent age, the late Russell Bulgin said the Nissan Micra was about as fascinating as a microwave oven. Russell, I wish you were here now. Compared with some electric cars, the Micra is like a Jay Gatsby party mingled with Rubens' *Rape of the Sabine Women.*

Years ago, before electric cars became the reality they are today, Renault's Patrick Le Quément was already speculating that a supportive soundtrack would be needed to make them amusing. BMW has that in its i8: the stepless power delivery has a synthetic backing chorus which imitates, if somewhat abstractly, the catastrophic internal events that make internal combustion more ferally exciting than magnetism. Somehow, this seems wrong. There are people in California advocating cool virtual sex with digital probes and wired full-body suits, but I think I am inclined to prefer the hot and messy original.

I have always loved the hot and messy sound of engines. Does anybody anywhere who has bought a Porsche with the *Sportauspuff not* switch it on?

Always? I grew up with a V8 rumble in the background and remember going to circuits and getting that prickly sensation in the pineal gland as you approached, when you could just begin to hear the yelps and howls of racing engines in the distance, long before you could see the cars themselves.

My father once owned a glorious curio: a 33rpm long-playing vinyl disc of the Ferrari test-driver, Mike Parkes, at Monza. In comedic contrast, there was Peter Ustinov's glorious recording of the fictional 'Grand Prix of Gibraltar' where he made all the commentary and engine noises himself. For those who find the deathly hush of today's Formula One dull, I recommend it.

In car design, noise tends to be a by-product rather than an end in itself, but I wonder if the coming acoustic crisis will force people to think about it a little more systematically. In 1916, the Futurist Luigi Russolo published a book called *L'Arte dei Rumori* (The Art of Noises). Of course, the Futurists were mad about machine guns and speed, but Russolo's argument was a very thoughtful one. Before the arrival of machinery, the world, save for the odd thunderclap, volcanic explosion or little passage of insistent birdsong, was … silent. Only when steam hammers, trains, drills and petrol engines came along was Nature augmented, or some would say defeated, by a mechanical symphony. Russolo thought this raised the game for composers who, in his view, would have to work very hard to make a more artistic noise than, say, a howitzer.

As I write this, I am thinking about the pretty mechanical noises that we will soon lose or have already lost. The whine of a Porsche fan and the clatter of valves or the wheeze of a carburettor. Of course, noise is a form of pollution and chastening evidence of wasted energy. Noise is aggressive and, etymologically, perhaps derives from the Latin *nausea*: it sickens and is aggressive. But it is also engaging and expressive.

Tesla was, himself, out there in the blurred margins of sanity. He said his eyes had turned blue because his brain was used so much. He was obsessed with a 'death ray' that would transmit electricity and believed that flexing your toes increased the IQ. Additionally, he kept pigeons in the New York hotel room where he led a life of well-publicised chastity. Would you buy a car named after this man?

MOTORSHOWS

Hitler and Porsche at the Fallersleben factory, 1938

Ferdinand Porsche was an independent design
consultant, long before his name became attached to
famous sports cars. His relationship with Hitler was
ambivalent, although he had no scruples in designing the
Silberpfeil (Silver Arrows) racing cars for Mercedes-Benz
and Auto Union as propaganda tools … nor in designing
the Wehrmacht's Nazi tanks and the Luftwaffe's V-1
buzz-bomb. But the greatest success of the Porsche
Konstruktionsbüro für Motoren-Fahrzeug-Luftfahrzeug
und Wasserfahrzeug was the Volkswagen, the Fuhrer's
Strength-through-Joy car.

General Motors Motorama, 1953

Motor shows became one of the quasi-religious events
where the rites of the automobile were celebrated.
General Motors' Motorama ran from 1949 to 1961, almost
exactly matching Harley Earl's hegemony as Detroit's
chief wizard of design. Here, before television became
the dominant medium, a fascinated public could gasp at
'Dream Cars' in this nomadic circus of kitsch. The 1954
Firebird – a crazy jet-powered nonsense – was, for
example, presented on a giant velvet cushion under
cinema lights. In this way, several generations had the
coordinates of The American Dream mapped for them.

The London Motor Show, 1966

It does no honour to the deep cultural history of the Ford
Cortina that it became known as the 'Dagenham Dustbin'.
It was more complicated than that. When the Edsel failed
so calamitously in the US, its designer, Roy Brown, was
sent to the Siberia that was Ford of Britain's Essex
operation. The result was the Cortina; commercially
speaking, perhaps the most successful British car dynasty
ever. Even its name was significant: in the early days of
package holidays, it evoked a Dolomites ski resort. John
Betjeman, the Poet Laureate, paid tribute: 'I am a young
executive. No cuffs than mine are cleaner; I have a
Slimline briefcase and I use the firm's Cortina'. Seen here
is a Mark II, designed by Roy Haynes, with a soft top
conversion by Crayford.

LOOSE LOCK NUTS

One Sunday I was invited to a convivial breakfast in Belgravia with some car types. Obviously, a good idea, but then the host good-naturedly explained that my 2016 nine-speed car with paddle shifts, voice recognition, collision avoidance system and driver-assists to keep you alert, even after several bottles of Morey-Saint-Denis, would not, if we were honest, be welcome. So, would I park it around the corner and join the others as they scattered brioche crumbs over the lovingly assembled collection of Cobras and E-types?

I decided that my problem was sourced in motor racing. Once upon a time, motor racing used to produce sporting heroes with a synoptic genius for expressing the human dilemma. I love it, for example, that Mario Andretti once said, 'If everything seems under control, you are simply not going fast enough'. That's a perfect way of expressing our common absurdity. Once, the fundamental risks racing drivers took inspired them.

And motor racing is the only way a classic is created. Long after its competition successes in the mid-fifties, Jaguar customers were still citing 'Le Mans' as a justification for buying the cars, even when they were being assembled to North Korean standards and finished in paint that looked like lab samples from a dysentery hospital. No matter how good its cars become, Lexus will never be a classic because it has never raced. In this sense, Ferrari, Porsche and even Jaguar will forever be untouchable. Lamborghini has trouble with a clear brand proposition, unless you are a teenage Emirati playboy, because it has never won a proper competition.

Motor racing was once a school of hard knocks (and, I suppose, NOx as well). This came powerfully back to me as I read some old, foxed articles about the speed-record attempts that Lotus made at Monza between September and December 1956. The period flavour was as delicious and as inimitable as a thriller by Eric Ambler or a horror story by Dennis Wheatley. You can imagine the sulphurous fog and the brown furniture, with a hint of rock'n'roll in the distance. From a garage in Hornsey, Lotus's chief mechanic, Mike Costin, organised attempts on the 750cc and 1100cc classes in a beautiful Lotus Eleven with an aerodynamic bubble canopy, lapped into the groove of the new head fairing, which he had designed. It was given a special paint job, they polished the undertray, the panel gaps were covered with tape for better aerodynamics (apparently decreasing the lap times by one and a half seconds) and the front brakes removed to reduce unsprung weight.

While today, Lewis and Nico may get stressed if their yoga class is late, their facials are delayed or their NetJet is not a current model Gulfstream, Costin drove the Lotus transporter across Europe. Before they had got to Seven Sisters Road, they had run out of fuel in one tank and were sucking air in the other. On the way to Folkestone, they ran over a petrol can which damaged the bodywork and a wheel. As they boarded the ferry, the float bowl fell off the transporter's carburettor. Still, with the brakes grabbing and the rear axle lock nuts working loose, they drove overnight from Dunkirk to Briançon, a distance of 1,000 kilometres.

Eventually, Costin arrived at Monza two days after leaving north London. His driver was Cliff Allison, although Lotus had made earlier attempts at the same record two months before with Mac Fraser and Stirling Moss, the latter's first drive in a Lotus. Moss had driven the car there behind a Standard Vanguard van. Fraser found the Monza vibration so damaging that he spent several subsequent weeks peeing blood. During Moss's attempts, in which he took sweets on board to sustain him, the canopy blew off, he was pelted with redundant rivets and eventually the whole rear section of the little Lotus blew off into the Monza undergrowth.

The very same thing happened to Allison, suggesting, if you ask me, a fundamental design fault, but 'lap times were not greatly affected', which tells you all you need to know about the black art of aerodynamics. Allison also suffered split manifolds (soon re-machined by a local workshop), the supercharger belts burning out and a misfiring engine. Still, they won all the FIA's Class G records. So Costin then drove back over the Alps at 12mph in a snowstorm and suffered broken shackle pins and two burst tyres. Hornsey perhaps never seemed so beautiful.

What can the relationship be between these harsh circumstances and the existence of the most beautiful of all cars? It's a curiosity of human nature that difficulty produces excellence. This is why London, while absolutely bloody, dangerous, dirty, expensive and dangerous, is a wonderful place. It is why racing cars of the mid-fifties were the best. Danger and hardship are to be encouraged.

And while no one wants to encourage carnage, the fact that Lewis and Nico might soon be in cars with canopies is a retrograde step for a sport that is already too sanitised, politicised and bureaucratised. And it is all connected to the reason why my techno-marvel with its nine-speed paddle shift is not welcome at a Belgravia breakfast with real sports cars.

ROME

True, the *ludi circenses*, the chariot races with their confident expectation of bloody and violent disaster, laid a cultural foundation for Roman driving. And there are times when circulating the Grande Raccordo Anulare – so much more poetic than 'ring road' – can seem positively gladiatorial. But really, there can be few major cities less suited to the car than Rome. And fewer still where cars are so much a part of the city's story.

I was there once for a dinner at the Circolo della Caccia in the Palazzo Borghese, a club of such ancient provenance and grandeur that it makes White's look like Center Parcs. My lugubrious neighbour, sadly pushing a slice of *vitello tonnato* around his antique plate, explained that he hardly ever drove into the city any more since the congestion was terminal, or worse, and in any case there was nowhere to park. Not that it was merely *molto difficile* to park, but that it was predictably impossible.

This was anticipated in the cinema more than fifty years ago: the opening sequence of Federico Fellini's *8½* shows a suffocating traffic jam as a means of suggesting the tense drama to come. And in Fellini's great masterpiece, *La Dolce Vita*, the cars also suggest psychological states: a race between a Giulietta and a Thunderbird is as profound as a contest between good and evil represented in a medieval fresco. Roman novelists understood the car too. At the beginning of *Il Disprezzo* (*Contempt*), Alberto Moravia describes the anguish caused by having to support a wife and a car, evidently two of life's burdens.

I spent hours in Rome leaning against the bar of the Caffè Canova, Fellini's favourite haunt, because it was in the shade. They have a loop of his films playing continuously and you marvel at the sight of a Lancia Aurelia B24 motoring very slowly, in black and white, across an absolutely deserted Piazza del Popolo. The beauty of the car, the strange emptiness of the space, the uncertainty of the destination: the sequence looks like no less than an emblem of Dante's road of life. Especially after six viewings, several glasses of Frascati and a bowl of green olives. That same lovely piazza today has Asian pedlars selling dayglo silly putty and counterfeit iPhones.

I mused on this decline and fall as I picked my way through the mess of Toyota Auris wagons on the via del Babuino which have replaced Fellini-era Fiat 1500s as the Roman taxi. And then I found a 1970 Rolls-Royce Corniche parked outside Rocco Forte's Hotel de Russie. It was olive green metallic with cream hide as battered as a rhino's scrotum and had English plates. It looked perfectly at home. There was a Panama hat on the rear shelf: the sight was like the synopsis of a short story. Twenty-minutes' walk later, I found a fabulous cream-coloured Fiat 2300 Coupé outside the Rococo church of Sant'Andrea della Valle. The fit was perfect.

Crossing the river by the Ponte Sisto, I walked into Trastevere looking for dinner and instead found a 1955 Fiat 500C Belvedere parked in front of a heaving, noisy trattoria. For me, almost nothing could be more perfect than this delightful little machine, an 'estate car' version of the last Cinquecento before Dante Giacosa's epochal 1957 Nuova Cinquecento. This one was in two-tone green and spoke of charming convenience and *l'arte difficile di essere semplici*. I checked and, four-up with some sparmannia cuttings in the back, it could reach almost 50mph. This was felicity.

And soon after, returning to the Caffè Canova, I read the news today, oh boy. FCA, which sounds like a dreary provincial firm of accountants, not Italy's premier manufacturer, was considering moving Fiat manufacturing out of Italy into low-cost Poland (where the current 500s are made). This to free-up production space for high-value premium products.

And who would notice? Because, very sadly, no one cares much about Fiat any more. Least of all, the Italians. Two generations of, 500 apart, stultifying product mediocrity has nearly eradicated a century of patiently accumulated affection founded in the customer's appreciation of innovation, style and charm. What question did the Tipo answer? Even Fiat did not know. It was culled after years of infectious apathy.

The 't' in Fiat, stands, of course, for Torino, the city that created grissini: a unique territorial sense is inherent in the company's name. But Fiat is now reduced to a single letter in the FCA acronym. And FCA is a faceless, deracinated financial vehicle. Debt reduction, not vision, is the corporate preoccupation. Shakespeare spoke of the folly of selling cheap what is most dear. And what is most dear about Fiat is its association with Italy, the land of Giacosa, Lampredi, Jano, Farina, Ghia, Bertone, Giugiaro and all those other resonant names belonging to bold, singular engineer-designers who made, for a while, *la macchina italiana* so distinctive and so alluring.

Genius loci was another idea donated to civilisation by the Romans. This sense of place, this sense of identity and belonging is extremely precious and will become more so in a heartlessly digitalised and drearily globalised world. In terms of brand value, a Fiat made in Silesian Tychy will be worthless. Whoever wants Polish pasta?

SOME UGLY VEHICLE

Will there ever be a beautiful SUV? Perhaps not. The name itself is revealing. There is a settled opinion today that the abbreviation stands for 'Sports Utility Vehicle', but pedants with historical perspective may prefer the less exalting 'Suburban Utility Vehicle'.

The first of the type was the 1935 Chevrolet Carryall Suburban, a military half-ton truck which had grown a car-like body the better to serve ruggedised users on their commute to the Westchester railway station. Of course, trucks have their place in the automobile pantheon (I am especially fond of the lovely 1950 Officine Meccaniche Leoncino), but suburbia perhaps does not. A Sport UV suggests exhilarating glamour and danger while a Suburban UV suggests deadening routine and tedium. The Cresta Run at St Moritz or the school run in Carshalton: which excites you more?

The question of whether or not there should ever be a Ferrari SUV may have been the source of the famous Vesuvian eruption at the company. The populist Marchionne perhaps said yes, while the elitist di Montezemolo perhaps said no. Here was a point of principle which, as the world now knows, the principled party lost so a Ferrari SUV is now a reality. (Although some would say that with the cumbersome FF, Ferrari was there already.)

Elegance seems not to be a possibility for the SUV designer. With its heritage in agricultural equipment, Lamborghini was well-placed to realise any potential beauty in the genre. Yet its LM002 was comedically ugly, gross, crudely detailed and ill-proportioned. More recently, the Bentley SUV has been frightening sensitive people. I won't call it concrete, rather 'weighty', evidence that Bugatti's jest that Bentley made fast lorries was one of his many true words.

Maybe the dimensional specifics of the SUV genre don't allow for that mysterious alchemy which turns arrangements of a car's lines and shapes into something beautiful. Mid-engine cars may be ergonomic horror stories, but they help the designer with their fundamentally appealing proportions. The height and ground clearance of the SUV will, surely, always make these cars look ungainly.

Factor in the givens about entry and departure angles and it means a lot of automobile stuff best left private is often exposed. Recently, in a low-slung car, I followed a new Range Rover and was mesmerised by how much was visibly going on underneath. An image kept on coming to mind of a plump Cruikshank-era servant who had hitched-up her skirt to bend over a steaming cauldron. Fascinating, but not beautiful. Then, if used as intended, an SUV will be covered with mud. Art and mud do not mix well.

What are the greatest SUVs? In 1946, the industrial designer, Brooks Stevens, began house-training Willys Jeeps: they grew cabins and became cute. Stevens was a fascinating figure whose work can be read like a psychographic of the American consumer's tormented imagination, with its unwholesome lusts and costive puritanism. In 1949, he styled the Harley-Davidson Hydra-Glide and soon after decided that absolutely the best idea in the world was to put a window into washing machines, so the chore of laundry could be turned into a visually exciting narrative for people dumb enough to stare at washing machines.

Then, seven years before the Range Rover, Stevens produced the Jeep Wagoneer, the first SUV in our sense. It had a big V8 and, with various comfort features, stood apart from rescued military vehicles, whose bombproof crudity and juddering progress had hitherto defined the type. We cannot say if the Wagoneer influenced David Bache to draw the Range Rover because no one knows exactly what inspired this enigmatic man to produce so ineffable a design.

Bache, I greatly admired, but it must be said that he lived in West Midlands suburbia and wore Cuban-heeled boots. Yet from this unpromising milieu and with this alarming wardrobe he produced a car that was more German than the Germans in the matter of *gute Form*. In anybody's list of the top ten cars of all time, the 1970 Range Rover would surely be included. It was almost perfect in every way a designer would define perfection. But it was not beautiful.

Maybe because Porsche has more of a historic commitment to extreme functionalism than to gracious beauty, it has had the most success of any pretender to the SUV kingdom. The first-generation Cayenne was a shocker, artistically, but, against howls of protest from aesthetes, soon gained acceptance. The second-generation Cayenne became much less difficult to look at, but is not a distinguished design: aesthetically, it is more an absence of negatives than a presence of positives.

And now there is the Macan. I have been driving a Macan Turbo, and while I am no connoisseur of vehicle dynamics, it seems to possess all the glorious precision and force of a Porsche sports car within a sensible, polite and utilitarian package. What a fantastic car. It's a brilliant compromise, but still not beautiful. Maybe there is an essential truth here: beauty can never be compromised. Nor, perhaps, suburban.

SYMPATHETIC MAGIC

There is no better evidence that marketing is an imprecise science than the Toyota Hilux. Who has not been impressed by news pictures of forty-three white pick-ups in a murderous conga-line on the Syria-Iraq border? It's fascinating that ISIS takes such pride in ownership, pride that would not be out of place in Westchester or Surbiton where radical Islam does not flourish.

Even more fascinating that someone is art-directing these propaganda images with all the skill of an infidel adman more used to organising car shoots on the Grande Corniche or that famous wiggly road in Tuscany they all like to use.

I was so fascinated and disturbed by ISIS's choice of vehicle that I asked Toyota for a reaction. Are you at all worried, I said, that prehistoric psychopaths identify with your product? Is it to any extent damaging your brand values?

And it is not only ISIS. Somali pirates also have a strong preference for the sturdy Hilux. It's the ride-of-choice for lawless killers everywhere. These questions were fielded boldly by Toyota. While properly disdaining associations with bigotry and savagery, the company can take some justified pride in the fact that its reputation for reliability is understood and appreciated by customers everywhere. Even by those who want to retrofit a Soviet-era DShK heavy machine gun on to the flatbed.

Indeed, interrogated on the matter of utility and robustness by a diligent reporter from the *New York Times*, a US Army Ranger said the civilian Toyota 'sure kicks the Hell out of a Humvee' (referring to the Army's clumsy AM General High Mobility Multipurpose Wheeled Vehicle). It was presumably this same fabled ability to kick the hell out of things that endeared the saintly Pope Francis to the Hilux he used on his South American tour. The Hilux congregation is a broad church.

As I say, when it comes to understanding the appeal that certain products have to specific groups of people, conventional marketing has little of value to teach us. You'd be better off reading Sir James Frazer, the pioneer anthropologist, whose 1890 masterpiece, *The Golden Bough*, is one of the richest sources of deconstructed magic and myth. Frazer gives us the concept of 'Sympathetic Magic' which is based on notions of similarity and contagion or, as he put it: 'like produces like … an effect resembles its cause'. Or put it this way: you are your car.

One of the richest subtexts in the narrative of the automobile is how certain cars became identified with certain people. In this process, each reinforces the other's brand values in a reciprocal arrangement as efficient and inevitable as a spinning crankshaft. Take, for example, Facel Vega. This company had no real credentials. Facel is an acronym for Forges et Ateliers de Constructions d'Eure-et-Loir, a metal-bashing business which made money from stamping fridge components. How sensible then to recruit Stirling Moss and Pablo Picasso as what we would today call 'brand ambassadors'. A stylish and brave racer and a prodigally talented artist, both successful with women! Who would not want to share that warm bath of positive attributes?

Or there was James Dean and his glamour of delinquency. He quickly identified Porsche as a car expressive of his beautifully disturbed character and duly died in a 550 Spyder when it hit, at speed, a humble Ford station wagon in the California desert. Porsche has since never been entirely able to shrug off a reputation for both speed and danger. By contrast, Jaguar built its reputation on the West Coast as the car favoured by Humphrey Bogart and Cary Grant. Each cruised Hollywood in an XK120. So that said elegant, but masculine. A decade later, the comedian Peter Sellers gentrified the Mini and John Lennon did the exact opposite with his psychedelically artworked Rolls-Royce Phantom V. Or there's John McVicar, the armed robber, and his special interest in hiding in the boot of a Ford Cortina. Suburbia had never been so exciting.

But the best ever annexation of human talent to machinery was BMW's Art Car programme which began in 1975. Here was Sympathetic Magic of a very high order. Commissioning Roy Lichtenstein, Alexander Calder, Frank Stella and Andy Warhol to paint 6 Series coupés was an inspiration. Suddenly, a German car acquired the values of a Midtown Manhattan art gallery. Customers could suspend their covert anxieties about compression ratios and, instead, imagine that their choice of car bought them entry to the private-view world of warm white wine, salted nuts and exhausted, intelligent chatter.

Never mind that at the same time, Germany's Red Army Faction was using BMWs as getaway cars after bloody heists. Indeed, there was amused talk that BMW might actually stand for 'Baader-Meinhof Wagen', the name of the Red Army Faction's celebrity gang. Like ISIS and the Pope with their beloved Toyota trucks, the art world and terrorist each used and enjoyed and identified with BMW cars. Was it art or violence that added more to the popular perception of BMW?

This is one of the great imponderables of aesthetics. Do the people make the car or does the car make the people?

I wrote that line, got into the car, pressed the button and was immediately stalled in traffic. The dream they sold me was a lie. But lies are soon forgotten and dreams last forever.

VALUE DENSITY

I have discovered the interesting, possibly priceless, concept of value density. It was put to me that a modern jet engine has one of the highest value-density ratios you can find, bettered perhaps only by sophisticated wristwatches. For every pound of weight, a jet-engine manufacturer can charge much more than, say, someone who makes sourdough bread or house bricks.

Indeed, the author of the article where I found this proposition, in *Eureka!* – the magazine of engineering design – had a gastronomic metaphor of his own. While Rolls-Royce's Trent XWB turbofan has a value density as high as they come, the Ford Fiesta is comparable in these terms to a Big Mac. Personally, I think that's a bit unfair because the Fiesta is an excellent little car and I'd be more likely to compare it to a prime fillet of wagyu beef than a greaseburger, but an important point has been made.

It is amusing how limited are the criteria we use to assess fine cars. We don't even use all the senses. When did you last hear anyone talk about the texture of, for example, the engine-turned instrument panel on a Bugatti Brescia? And the smell of a car is usually ignored. True, Ferrari's Luca Cordero is fond of quoting the Paolo Conte song which says a sports car should smell of paint and sex, but in general olfactory criteria are ignored in the design and the appreciation of cars. Taste too. Steve Jobs of Apple once said, 'you know a design is good when you want to lick it', but I have never seen that principle tested on, for example, a race-hot D-type.

Which brings me back to value density and the question of weight. This is much too rarely considered, yet an essential factor in excellent design. The disciplines involved in reducing weight not only contribute to better performance, but generally force an aesthetic elegance on the design of components. Any fool could make, shall we say, a suspension wishbone that was indestructibly durable and weighed 200lbs, but to make that component just as functional while reducing the weight by ninety per cent takes a sort of genius. And the results of such economy are often beautiful and ingenious.

This was why Gordon Murray's original McLaren F1 was so enthralling. With the uncompromising obsessiveness that comes from the freedom to exceed an unlimited budget, Gordon commissioned ultra-lightweight, custom-made titanium nuts. The most desirable new car of recent years is surely the Alfa Romeo 4C whose very essence is a *leggerita* that leads to a whole lot of *bella figura*. Colin Chapman was messianic about weight. One of his best nostrums was 'simplify and add lightness'. One of the reasons why the Lotus Seven remains essentially unchanged over more than sixty years is that Chapman's mania for lightweight forced an unimprovable simplicity on the car. Like any great design, there is nothing that needs to be added to or subtracted from a Seven in order to improve it. (Except, of course, a better convertible top.)

The pursuit of lightness is a stimulus in other design disciplines. When the prodigious and ever so slightly blowhard Buckminster Fuller met the architect Norman Foster, he asked: 'How much does your building weigh?' Perhaps concepts of value density apply in the inverse to architecture, Canterbury Cathedral being both heavier and more valuable than a Portakabin, but it was a question that made Foster redouble his own efforts to achieve lightweight efficiency in his buildings.

These thoughts came crashing together when I was a judge of the Schloss Bensberg Classics at Bergisch Gladbach, Cologne's Virginia Water. This is a concours d'elegance intended one day to rival Pebble Beach and the Villa d'Este, but perhaps only when the Rheinisch-Bergischer Kreis enjoys the same weather as California and becomes as beautiful as Lake Como. Here, in moments when the conversation of the other judges exceeded my grasp of the subject, I started formulating my own concept of elegance: poise, wit, grace, gentleness … good manners. But most of all: a lightness of touch.

But how rarely is lightness of touch considered valuable? True, as at any world-class concours, there were some splendid nutters. I especially enjoyed the man who had rebuilt – I perhaps mean re-imagined – his Cisitalia starting only with a pair of carburettors. True, some winners at this concours d'elegance had lightness in their spirit: a 1954 OSCA 2000S or a 1952 Siata Daina 140S, for example. But considerations of lightness played no part in the judges' decision to give awards to a 1933 Duesenberg SJ, a car as a big as the Ritz, or a 1963 Studebaker Avanti.

And then there was the Best of Show. The public chose a delicate 1955 Porsche 356 Pre-A Speedster, while the judges chose a complex and baroque 1957 Maserati 150 GT. Chairman, Franz-Josef Paefgen, said, a little disdainfully, 'the public does not like exotics'. At this moment I suddenly thought of Kate Moss who said that nothing tastes as good as being thin feels. The Maserati's owner confided in me that his car was an absolute dog to drive, while the Porsche was and remains a thing of dynamic wonder. We need to do a value-density check to establish the true meaning of elegance.

BEAUTY

Have we reached the end of beauty's road? Forgive me if I mention it too frequently, but I often think about Albert Camus's crash, because all the poetry and bathos, the romance and absurdity, the glory and calamity of the motor car are contained in its harrowing details. Camus's last words to the driver were, reportedly: 'What's the hurry, little friend?' What, indeed.

Camus said and wrote many other haunting things. Especially on beauty. So far from finding it enjoyable, he found beauty unbearable because it 'drives us to despair, offering for a minute the glimpse of an eternity that we should like to stretch out over the whole of time'. That beautiful clear-eyed and floppy-haired youth will one day be a doddering, capillary-busted, malodorous old man. (Sound of squealing rubber and tearing metal.)

Beauty, however troublesome to define, has been an agreed pursuit of Western civilisation since the Greeks. But it has run its course. About twenty years ago, ugly cars began to appear. I don't mean maladroit and ill-considered dogs: products of ignorance, cack-handedness and cynicism. I mean designs intended to confront and disturb consumers, rather than to seduce and pacify them. Note, perhaps, that our word 'ugly' comes from the old Norse *ugga*, which means aggressive. That Huracán is ugly, if you ask me.

But there had long been speculation about uglification even before Chris Bangle's notorious 2003 BMW 5 Series made beauty historic. Alice, for example, was having a conversation with those two fabulous creatures, the Gryphon and the Mock Turtle, in one of her adventures in Wonderland. The Gryphon says to her: 'Never heard of uglifying! You know what to beautify is, I suppose?' Alice replies: 'Yes, it means to make something prettier'. So, the Gryphon goes: 'Well, then, if you don't know what uglify is, you are a simpleton'.

Cars were once named after pretty creatures, Impalas and Jaguars, or Panthers, for example. Yes, I know we have had Volkswagen Beetles and Rabbits too, but they are rule-proving exceptions. People called the Bangle 5 a 'slug' and even in these days of Postmodern relativism and all-species inclusiveness, no one finds slugs beautiful. But Bangle was not wrong. As a hysterical Victorian lady novelist said, 'Familiarity is a magician that is cruel to beauty, but kind to ugliness'. Familiarity has been kind to the 5 Series: all cars in its category now look rather similar. But I said 'similar', not 'beautiful'.

More evidence that beauty has run its course? At Apple, Jony Ive took physical perfection to its limits. Aesthetically, there is nowhere else to go: an iPad Air is the *reductio ad absurdum* of the Bauhaus aesthetic. There's nothing to add and nothing to take away. And very beautiful it is too, but Apple is a stock-watching mega-corporation and when sales of perfection decline, as they will, perfection will not be good enough and the designers will be called to uglify.

This used to be called planned obsolescence. First evidence? The coloured iPhone 5c. I haven't spoken to Jony for a while, but a dayglo iPhone had a suggestion of a defeat in an over-my-dead-body battle. There was talk once of Ive creating an Apple iCar, a more aesthetically disciplined smart, perhaps, but that's probably never going to happen.

But were there ever actually any standards of beauty? The critical language of car design uses tropes such as 'muscle under the skin' and 'feline stance' and 'subtle aggression'. Certainly, some things are known about the traditional sources of automobile beauty. The proportion and articulation of the big felids, those jaguars and panthers, is positively inspiring. So too is the sense of penetrative thrust of a racing yacht. A beautiful car might also imitate the athleticism of a mid-size African antelope.

But these reference points are ancient. Maybe we are losing a sense of beauty because no one really knows what a car is for any more. When was the last time you packed your pigskin suitcases into the boot of your wire-wheeled Gran Turismo and motored to Positano? Art shows the way. Visit the terrifying Hong Kong Art Fair and the experience makes the polite history of art from Giotto to Pollock, via Picasso, seem quaint. Art isn't art any more, it's a new financial asset class. Is there a whole new aesthetic coming up over the horizon, or are we truly done with ideals of beauty?

The Chinese are very busy, if misguidedly, acquiring Western 'positional goods' (many, I am told, think Gucci is British) and buying Bentleys, Porsches, Ferraris and Rolls-Royces. This provides a fragile competitive status in our globalised scrum, but what's going to happen when every single citizen in Shenzhen wears horsebit loafers and drives a Continental Supersport? What next? I wonder if Western notions of beauty and taste can survive the decline of Western political and economic authority.

Meanwhile, if any designers know what beauty is, they are keeping it to themselves. Or have you not seen a BMW X1? Camus apart, the best observer of beauty is Harvard professor Elaine Scarry who holds the sonorously named Chair of Aesthetics and the General Theory of Value. Professor Scarry's general theory says the best test for beauty is: do we want more of it? My initial reaction is 'yes', but then I start to wonder.

WIDTH

Anybody who has reversed a Lamborghini into a tight parking slot knows at least two things. One, the exquisite pain caused by the sound of expensive alloy grating an unforgiving kerb. This, in my own case, is kept fresh in the memory by recurrent nightmares. Two, the sovereign importance of visibility when it comes to controlling a car.

The need to look over your shoulder to detect danger, or kerbs, where they are not the same thing, is instinctive and prehistoric. It's an unlearnt response to a crisis, an intelligent survival characteristic. Owls, the wisest of creatures, can turn their heads through 270 degrees. But in a Lamborghini, or practically any mid-engine car, the oppressive architecture does not allow you to turn your head with advantage. Sure, you can turn it, you just can't see anything when you do.

So, if parking is a nightmare, the open road is even worse. Granted, few things are likely to be coming up behind a fast-moving Lamborghini, but when overtaking I always look back to check, a sort of *à la recherche de la route perdue*. I cannot stop doing it. And being thwarted in retrovision increases panic and diminishes confidence. No amount of mirrors, cameras, blind-spot or traffic-alert systems can compensate for a good, clear vista.

This is just one of many, many reasons why the Giugiaro Fiat Panda was a much better design than the current Lamborghini Aventador. You sit upright with non-distorting flat glass everywhere and with so little power to dispense, you can drive fearlessly on or over its modest limits. And the other important performance characteristic of the 652cc two-cylinder ur-Panda is that it was, at 1.46 metres, wonderfully narrow. The hideous Aventador is 2.03 metres.

These thoughts about visibility and width were with me on a week spent going up and down the busy *autostrada* between Bari and Brindisi. In a rare concession to Health & Safety, Italian *autostradisti* do now drive, as law demands, with the lights on as if illumination alone might assist collision avoidance. On the whole, it seems to work.

And you realise too that Italian motorways were designed in the days when a Fiat 1100, with the proportions of a coffin, was a big car. Personally, I find narrow lanes existentially harrowing even if there are sound psychological reasons for them: David Shinar, a traffic engineer at Israel's Ben-Gurion University, says wide, comfortable roads with good visibility are dangerous because they encourage excess. Who wrote the rule that humans must be rational?

Historically speaking, cars are getting wider. Designers enjoy width because it allows them more latitude with proportions and creates dramatic effects. Look carefully at Art Fitzpatrick's glorious airbrush renderings for Pontiac in the sixties and you'll see that he exaggerates the width by about one third: the pictures are both ridiculous and impressive, as was intended. These Pontiacs were already what Americans endearingly called 'full size' (to distinguish them from the despised new compacts). This meant six passengers in total. If manufactured, an Art Fitzpatrick Pontiac could have (most amusingly) accommodated six abreast on each of two benches.

But wider cars are not better cars. The first Porsche 911 was such a joy because, like the Panda, it had flat sides. Its front wings which acted like gunsights allowed aiming to pleasantly accurate effect while at 1.7 metres the car was precisely narrow, like a fast scull. We can agree that a new, wide-body Porsche is technically superior in every way, but width inflation causes nagging user anxiety.

The latest Ferraris are visually sensational, but their enormous width disqualifies them from everyday use, certainly in Italy. One of the loveliest roads I know runs through the olives and trulli of the Valle d'Itria between Cisternino and Martina Franca. A new GTC4 Lusso at 1.98 metres would simply not fit, while at 1.7 metres the exquisite 1969 Dino 246 would be perfect. This most beautiful car was only 38cm wider than a tiny Nuova Cinquecento. Talking of great beauty, a Jaguar E-type is even narrower than the Dino at 1.66 metres.

The resolute pursuit of width and lowness was one of the many bizarre aesthetic adventures inspired by that great wizard of kitsch, Harley Earl. Every new model year during his reign at Detroit, GM's cars got wider and lower, and longer too, with absurd results. One year he realised that the Chevrolet Nomad wagon was now so low, the roof had become visible for the first time. He could not abide an undecorated space, so grooved it.

There is something decadent about ever-increasing width, although decadence is not always a bad thing where art's concerned. But how odd in the Darwinian sense. Ever-wider cars look ever more impressive, even as they become ever less useful. I doubt there is time between now and the moment when cars become illegal for this unstoppable process of engorgement to be reversed. I'll just leave you with a solitary thought. The Mini was the cleverest car design ever and it was 1.41 metres wide. The Ford Edsel at 2.02 metres, the most stupid.

A DISCREET BUTTON

We aesthetes are not necessarily attracted to speed. On the contrary, having the world pass by in a rapid blur that cannot be processed is a bit of a problem if you like looking at things. Staring is preferable to blinking. But speed does have its aesthetic aspects. As Aldous Huxley observed, speed was the sole novel sensation available in the twentieth century.

And there is endocrinology to consider as well. Acceleration and speed excite glands whose secretions change, perhaps enhance, our perceptions. Meanwhile, danger is itself a stimulus which sharpens the reactions. Speed thrills.

Hunter S Thompson, author of *Fear and Loathing in Las Vegas*, variously rode – in states of high intoxication – a Vincent Black Shadow, a Bultaco Matador and a Ducati 900 SuperSport. Before shooting himself and then having his ashes shot into space on a rocket released by Johnny Depp, he wrote: 'Faster, faster, faster until the thrill of speed overcomes the fear of death'.

Quite so, but speed is not an option available to urban aesthetes. In London, traffic now moves more slowly than it did in the days when the major source of pollution was horse manure. And I believe aesthetes are, essentially, urban. As anyone who has visited desolate, lifeless Poundbury or the ludicrous, posturing Soho Farmhouse knows, the creative spirit withers when surrounded by too much greenery.

But travelling slowly, especially on foot, my urban habit, allows plenty of time to stop and stare. And one of the things I think as I pick my way through queues of enraged, thwarted drivers is that never have cars been more artistically various and technically competent. Never have they been more practically useless.

On my footsore daily commute, I pass a house in Belgravia whose off-street space is occupied by a BMW i3 the colour of a laboratory worksurface and a Lamborghini Aventador the colour of an irradiated macaw. Marvellous machines, each. And then I think of the glum Søren Kierkegaard who thought 'the best demonstration of the misery of existence is by contemplation of its marvels'. To get an idea of the Sublime, simply measure the distance between the dream of mobility the BMW and Lamborghini represent, and the matter-of-fact reality endured by their users.

But there are times when I do use a car: at night and at weekends. And the car I have driven most satisfyingly is the Audi SQ7. Aesthetes are neither emotionally inclined nor technically competent to do road tests, but, if we are (ever so briefly) discussing speed, let me say immediately that this huge car is one of the most sensationally, yet controllably, fast cars I have ever experienced. I can think of nothing I would prefer for a long journey.

However, my unlikely claim for this Audi is not about dynamics, rather aesthetics. Great art may reveal itself immediately, but truly great art takes time to appreciate. It will yield its pleasures only after long contemplation. At first, I thought the refreshed and squared-off Q7 was a maladroit calamity: both annoyingly reticent and ham-fistedly aggressive. But gradually, after stopping staring a while, I came to understand how subtle and well considered it is.

But best of all is the interior, something which can be enjoyed whether congested or liberated. And it's not a matter of style or innovation: there is little here to excite, no flamboyance, no extremities, no conversation pieces. In fact, the interior is conventional and conservative. But the pleasure comes from the immaculate execution. Proof here, if proof were ever needed, of Le Corbusier's belief that 'good design is intelligence made visible'.

In particular, I loved the third row of seats. And that is a sentence I never thought I would write. They are deployed by a discreet button in the load space, almost invisible except for a tiny, dull, glowing red lamp. Press the button (which has no lost motion) and a servo whirs them into place. To stow them, press again and they decline, the headrest snapping shut halfway through the travel. And when stowed, they form a perfectly flat floor, flush with the larger stowed seats in front. The action is absolutely beautiful and the result worthy of long, long contemplation.

This result speaks of clarity of thought, and methodical execution, together with a degree of care unusual in any business. The depth of implied expertise is profound. I have heard Range Rover owners complain of the folding mechanism of their passenger seats. Playing with the Audi's third row, I found myself mumbling: 'Come over here, Gerry (McGovern). Stop polishing your Church's loafers! Put your bloody Bulgari catalogue down and take a look at this!'

Here are precision and accuracy. But precision and accuracy are not quite the same thing. If interested, I would recommend checking their definitions. Why? Because I think in these troubled times, the greatest satisfaction from car use or ownership will come from philosophical speculations, not mashing the pedal to the metal. In an Audi SQ7 you can enjoy both.

RACERS

James Dean in his Porsche 550, Vine Street, 1955

Alec Guinness told James Dean: 'If you get in that car, you
will be found dead in it by this time next week'. Dean had
already sensed the car's danger: George Barris, a Los
Angeles hot-rod craftsman, painted the legend Little
Bastard on the Porsche's tail. On 30 September 1955,
Dean set off to the races at Salinas. In celebrity fashion,
the photographer Sanford H Roth, on assignment for
Colliers magazine, followed the Porsche in convoy. Dean
drove fast. Approaching the junction of Route 466 and
Route 41, near Cholame, a Ford Tudor saloon made to turn
left and crossed the centre line. Dean's Porsche was
travelling at 85mph when it hit the big Ford head-on.
He was dead at twenty-four. Roth took the gruesomely
memorable pictures of the crash scene.

Steve McQueen in his Jaguar XK-SS, Sunset Boulevard, 1963

Steve McQueen was only twenty-eight when he bought
this Jaguar XK-SS, in 1958. Posing here at a studio lot on
Sunset Boulevard, at the time having played only bit parts,
McQueen was better known as a brave motorbike rider
at Long Island City Raceway than a Hollywood celebrity.
The XK-SS was a house-trained version of the Le
Mans-winning D-type. McQueen's car was delivered in
white with red upholstery, but he changed it to British
racing green with black leather. He called this difficult car
Green Rat, in imitation of Dean's Little Bastard. Vehicles
often complemented McQueen's cinema roles: a Triumph
TR6 Trophy bike in *The Great Escape*, and the Boss
Mustang in *Bullitt*.

Stirling Moss competes in the Mille Miglia, 1955

Perhaps the greatest ever performance by a racing driver?
In 1955, Stirling Moss drove his Mercedes-Benz 300SLR
– a Grand Prix car modified for road use – at an average
speed of 97.96mph through the Italian countryside on a
thousand-mile circuit of Brescia-Rome-Brescia. Moss had
personal control as finely tuned as his car control. His style
as a driver was completely distinctive: the head cocked
slightly to one side, arms straight. Inimitable.

SHAKESPEARE CAN'T DO JIVE TALK

I looked around the handsome interior and wistfully thought about the beauty of it all, stroking the finely textured instrument binnacle, admiring the clear graphics of the instruments while pleasantly absorbing the general ambience of well-being. Then I turned on the ignition. Another half turn, and with a fine whine that spoke eloquently of very high technology, the starter motor engaged the flywheel. The engine promptly came alive. Like an animal stirring, the entire car powered up and we were good to go. I cautiously reversed out of the dark, cramped garage, anxious about the vitreous-finish paintwork, but was soon on a clear stretch of road in gloriously misty-sunny daylight.

On a good day, in the right mood, there are few experiences better than driving a great car in a romantic location. Snick, snick, snick I went through the gears and, while I am an unremarkable driver, since the road was empty, I had an enjoyable time attacking the corners on the epic coast road from Lourinhã to Peniche. You know the thing: juggling with the variables of braking points, turn-in, clipping apexes, gear selection. Feeling forces load up and then disperse in a series of exciting peaks and dips. Truly, a delicious sort of erotic intercourse with a machine. What was the car? A dirt-cheap Hyundai rental I had picked up from Lisbon's Portela Airport. I doubt I could have worked this road faster – or more comfortably – in a Porsche 917.

Like childbirth and bringing up children, no one ever tells you the raw truth about classic cars. It's a survival characteristic. If we were realistic about the pain of parturition and the fatigue (not to mention wince-inducing expense) involved in childcare and education, human reproduction would promptly cease. Civilisation would end. And with classic cars, if we spoke the truth, or, at least, faced the facts, we would be out of business. But we don't. We bash on. If we were cold-eyed realists, we would have Hyundais. We are, instead, romantics.

But my experience of driving or owning classic cars has been universally dismal. The fact that I am still enjoyably engaged with them is evidence of man's laughable folly, of the triumph of blind hope over cruel experience. I remember the little Fiat Nuova Cinquecento I bought for my wife. Ineffably cute, certainly, but it was like owning a sick pet: adorable, but tragic. It could not be made to move.

Or the Citroën DS I drove recently. This astonishing car inspired Roland Barthes's wonderful line about design being 'the best messenger of a world above that of nature'. Yes, maybe, but the awful creaking and wheezing and cumbersome demeanour of the fabled Goddess made me yearn for something brand new, precise and Korean.

But this, of course, has nothing to do with it. Complaining that (most) classic cars do not work well is like moaning that you can't put Sèvres porcelain in the dishwasher, that Rembrandt is low-res, Shakespeare can't do jive talk, Abbey Road has crude stereo separation, and Jack Kerouac took drugs. The whole point of any classic – in any medium or genre – is that it transcends the ordinary and defies rational criticism.

For more than thirty years I have been fielding questions about 'classic design'. I can't say I have a definitive answer, but I do have the response well sorted. Any classic has to have an ambiguous relationship with time: it must speak of the age that created it, but also be beyond the basic cycles of fashion. And classics must tell a story, evoking a mystique beyond the here-and-now. Additionally, they establish a type: true classics have neither precedents nor successors. They are magnificently singular. And, of course, desirable.

In fact, I think the absolute essence of a classic car is the way it excites desire. That's to say an anticipation of pleasures to come. Look at that Lancia B24 or that Lotus 15 and you start an imaginative, rather than a real, journey. In a sense, desire is the opposite of nostalgia, because nostalgia looks backwards while desire projects yearnings into the future. It does not matter a pinch of raccoon excrement if the experience of driving or owning a classic is often compromised, a classic speaks to a higher level of psychological engagement than that mondain journey from the first to the second letter of the alphabet.

As with people, flaws and mistakes make cars interesting. The baseball sage Yogi Berra said if the world were perfect, it wouldn't be. The Hyundai is perfect yet isn't desirable. The Thunderbird, Jaguar, Cinquecento and DS are comedically imperfect, but I want one of each. Sometimes I think the only certain thing about human preference is its total lack of rationality. Thank God. Otherwise, we really would all be in Hyundais.

PAGODA

In 1965, a US Patent was filed for a 'motor vehicle with a concave top'. That vehicle was the W113 Mercedes-Benz 230SL, introduced two years before and known ever since as the 'Pagoda' on account of an elegant depression in its roof.

To get a sense of the enduring aesthetic thrill of this magnificent car, simply consider its contemporaries now lost to rust, the crusher and memory. Hillman Imp? Triumph 2000? True, the 230SL shares a birthday with the original Lamborghini 350 GT, but Carrozzeria Touring's aesthetic is located in the fifties, while the Mercedes looks fresh nearly sixty years later. The Lamborghini is satin couture and valve technology; the German car is polyester and transistors.

But what have oriental shrines got to do with automobile hardtops? Concave curvatures achieve good strength-weight relationships in structures, as the architects of the Horyuji temple in Nara knew. But in 1963 strength and weight relationships were also a preoccupation of the safety-obsessed Mercedes.

The 230SL's designer – a Bordelais, called Paul Bracq – explained, speaking to-camera in a blazer and cravat, that he determined 10cm was the optimum measure of concavity when applied to a car's roof. Our impressions of cars are formed by minutiae, but never can a mere 10cm have contributed so much to perception and reputation. The W113 has many aesthetic and technical virtues, but that Pagoda roof made it exceptional. That Pagoda roof made history.

But oriental architecture is only one of the elements in the 230SL's complicated rootstock of meanings. Bracq's master was French industrial designer Philippe Charbonneaux, who had done time at GM's Tech Center, in Warren, Michigan, and before returning to France contributed to what became the gloriously vulgar '53 Corvette: Detroit kitsch at its most deranged. Back in France, Charbonneaux created the 1957 Teleavia television set, whose pod-like swivelling screen predicts Jony Ive's iMac. Charbonneaux, it may safely be assumed, had an ambition to become Paris's Raymond Loewy: a slick vulgarian adroit at crowd-pleasers.

Then there are the levels of meaning deposited by the historic owners of any car. John Lennon, flush with royalties from *Please Please Me*, drove a 230SL. So too did Sophia Loren. Priscilla Presley was given one by Elvis and that car is now in the shrine at Graceland. Audrey Hepburn drove a Pagoda in the 1966 movie, *Two for the Road*, and fourteen years later, Helen Mirren was accommodated in a 280SL Pagoda in the Bob Hoskins caper *The Long Good Friday*. It's fair to say that pop and glamour have played their part in our appreciation of this car.

When Paul Bracq arrived at Sindelfingen from Paris, in 1957, the presiding spirits of Mercedes-Benz design were Friedrich Geiger a 'test engineer in the styling department' and Karl Wilfert 'chief of car body development'. These men's lofty credentials were based on their contribution to the epochal 300SL of 1954. If cars can be read like books, the 300SL was a design encyclopaedia where every entry was a gem. The 230SL is its successor.

But even more influential than Geiger and Wilfert was Béla Barényi, one of the great Austro-Hungarian engineers, who had made Mercedes-Benz his workshop. A cultured man from a wealthy, opera-loving, officer-class family, Barényi's obsession was safety. He created the crumple zone, first employed on the tailfin W111 of 1959. He made wipers disappear behind a scuttle: a device which defines genius in engineering, at once clever, practical and beautiful.

But Barényi has another part in the Mercedes-Benz mythology. The company identifies him as the 'intellectual father' of the Volkswagen. Indeed, there are Barényi drawings from 1924-5 of a rear-engine-car with a distinctive 'beetle' profile. So that was the People's Car. And the 230SL Pagoda was the Beautiful People's Car. In 1970, Janis Joplin took part in a poetry jam in a New York bar and the result was the lyric, Oh Lord, won't you buy me a Mercedes-Benz? She died three days after it was recorded. The song may have been a counterculture satire on consumerism, but the beauty and desirability of the Pagoda cut through the fog in Joplin's drug-addled brain.

Meanwhile, Wilfert had wanted the car to express stability and visibility. Stability was an established Mercedes-Benz value, and visibility was becoming an ever more desirable attribute for consumers. It seems that the patent-heavy Barényi, interpreting Wilfert, suggested the pagoda roof to the cravat-wearing Bracq as a means of adding rigidity to an audaciously light and airy glasshouse.

Essentially, what Paul Bracq did was to bring ooh-la-la French style to the very straight Sindelfingen. There are sketches by Geiger which show the general arrangement of a car that must be a predecessor of the 230SL: it is lean – almost austere – linear and symmetrical in profile, with equal emphasis fore and aft. Bracq developed this conceit in voluptuous renderings which added elan and esprit, but any tendency to Charbonneaux excess and frivolity was ultimately contained by the strict discipline of Mercedes-Benz studio practice. The production car was clicked back several notches from Bracq's most uninhibited drawings.

The resulting Franco-German 230SL is a nearly perfect composition: symmetrical in several measures, disciplined and refined. It has a gentleness that could be called reticent and feminine were it not a shape which somehow also expresses the weight and integrity of the car's engineering. The slenderness somehow emphasises sturdy foundations. The character lines are wise and subtle, the stance secure, yet agile. Lights are pushed out to the corners, adding to the visual tension. There is no flab.

Nothing is done to excess, but nor does the 230SL look stripped and bare. On the contrary, it looks sophisticated and expensive. That effect is partly to do with the exquisite execution of the car's jewellery: the application of chrome stops only just short of being lavish and the perpendicular Bosch headlamps seem to express wisdom and enlightenment. And they also make the car look modern, at least in a rather sixties fashion. There is nothing that needs to be taken away and nothing that needs to be added. That's a definition of aesthetic excellence. It is a perfect expression of what it is.

Bracq also drew the colossal Mercedes-Benz 600, as pharaonic and gross as the Pagoda is aristocratic and spare, but nonetheless a car with a surprisingly graceful profile and, again, an exceptionally airy and open glasshouse. Then the designer, impatient, perhaps, with his slow ascent up the Mercedes-Benz hierarchy, returned to France and coachbuilders Brissonneau et Lotz. Here, with Jacques Cooper, a Raymond Loewy alumnus, he worked on the design of SNCF's Train à Grande Vitesse, the TGV.

In 1970, Bracq became design director of BMW. But in Munich, while he established a design language for the 3 Series and 5 Series, he was a professional calamity, a temperamental outsider: reluctant to contribute to a design process that is essentially collaborative, preferring to sequester himself in his own office, producing drawings in competition with his own staff. Ian Cameron, designer of the superb first BMW Rolls-Royce Phantom, put 'fame' in inverted commas when he wrote to me about Bracq. Although he did concede that Bracq's 1973 BMW Turbo concept with gull-wing doors was 'pretty amazing'.

As a result of his unbiddable personality, when Bracq left BMW for Peugeot, his appointment was only as Head of Interior Design. This was partly because of his reputation for being difficult, but also because Pininfarina held the principal Peugeot design account. Instead, Bracq busied himself with his apologia: a book called *Carrosserie Passion*, which was published in 1990.

I found a copy of this frankly hideous book in a brocante in the Flower Market in Nice, some years ago. The cover is a garish Bracq rendering of realised and unrealised designs, including something a little like a Mercedes-Benz 600, swooshing polychromatically hither and yon. Like an explosion in

Philippe Charbonneaux's rumpus room, the Bracq book demonstrates that exquisite taste in one design discipline does not readily translate into another.

Even if the interior of the long-forgotten Peugeot 604 was an educated take on French notions of luxe, no one can blame Paul Bracq for not equalling the 230SL. Someone once said to novelist Joseph Heller, 'You have never written anything as good as *Catch 22*'. Heller replied: 'True, but nor has anybody else'. A version of this line would be very useful to Paul Bracq.

The Germans are uneasy with luxury. Henri Racamier, founder of what has become the earth-girdling grande marque and cashmere conglomerate that is LVMH, once explained that French luxury is feminine: champagne, frocks and jewels. But, Racamier said, English luxury is masculine: whisky, guns and tweed. He had nothing to say about the Germans.

Instead, the German concept of quality is founded more on the structural principles that inform both architecture and engineering. It is a belief that a building or a product that responds intelligently to the Laws of Nature will, almost inevitably, be a thing of beauty. Of course, there is evidence today that the Germans, perhaps especially Mercedes-Benz, have rather loosened their grip on this fine belief system. But in 1963, it still held good. The architect Heinrich Tessenow believed, 'The simplest form is not always the best, but the best is always simple'.

And the Pagoda is, as Tina Turner said, rather later, simply the best.

SHOWROOMS

Like department stores, car showrooms will soon be things of the past. Today, the bricks'n'mortar version of retail looks as quaint as a pony and trap did when they discovered oil in Humboldt County. A Tesla, for example, can only be ordered online.

And long ago, some Japanese manufacturers had the salesmen make home visits, obviating the expense of occupying high-priced, empty premises and toying with their honourable, but mercury-poisoned, bento boxes while waiting solemnly for a Corolla customer to amble in.

But some of my own finest moments came in car showrooms. The bookends were these. First, I was very small, in 1959, when my father took me to see the original Mini. It was in a part of Liverpool known as The Rocket, on account of Stephenson's steam engine starting its heroic Manchester run thereabouts. Even a toddler could see that, like the puffing and hissing Rocket, the Mini was a marvel.

Fifty years later I was at a Tokyo party in an impossibly lush Lexus showroom (probably by then a 'brand awareness centre') when I assumed my slightly trembling and blurred vision could be explained by the very many bamboo-leaf martinis I had drunk, only to be told there was a *jishin* of about three on the Richter scale going on.

Somewhere in between I learnt that, in the sixties, Bernie Ecclestone operated a showroom in Bexleyheath where he sold MGBs to Lulu and Cilla Black, 'Hit Parade' women singers of the day. This same showroom used Modernist glazing, white tiles and dazzling spotlights. I still enjoy bringing the image to mind. One has thought many things of Bernie Ecclestone, but rarely has he been appreciated as a pioneer of feminist consumerism and Minimalist architecture.

And that image always makes me think how very little attention is given to how cars are accommodated when they are not moving. Given that the automobile is *the* most designed object we consume, it is remarkable how (relatively) impoverished is the history of car-related architecture.

Of course, there are notable exceptions. The *locus classicus* of the showroom will forever be the Park Avenue and 56th Street premises that Frank Lloyd Wright designed, in 1954, for Max Hoffman, the entrepreneur who introduced Jaguar and Porsche to the United States. Here, Wright essayed the astonishing helical ramp design which later formed the basis of his Guggenheim Museum. Artistically, it was an inspired resolution of the conflicting demands of Dan Dare and the Bauhaus.

And then there are parking garages, a low-brow, trash-culture novelty celebrated to great effect in Simon Henley's *The Architecture of Parking* (2007). The masterpiece here is Jean-Michel Wilmotte's 1994 Parc des Celestins, in Lyon: an amazing seven-storey cylinder which humbles the resident Peugeots. In the same year, the Winterthur architect, Peter Kunz, began work on his Garagenatelier in Herdern, Switzerland: five concrete cubes, whose front elevation is all glass, are set into a gentle, grassy hillside. For a client who wanted to show off his collection, Kunz built something which beautifully equivocates between garage and site-specific installation.

The larger issue here is that as cars, especially interesting cars, become ever more difficult to enjoy on public roads, people will enjoy them more and more for their aesthetic content than their dynamic potential.

Thus, 'sky garages' are becoming quite the thing where HNWIs cluster at altitude. At the Hamilton Scotts condominium in Singapore, residents are offered an 'en-suite sky garage' as they once might have been offered a second loo, or a fitted kitchen, or a waste-disposal unit.

A helpful explanatory video from the developer explains the concept. You arrive at ground-floor level in a Ferrari or a Lamborghini. A glass-walled lift takes your car up to your fortieth-floor apartment and, when it comes to rest, you contemplate your 599 against a hazy backdrop of the Singapore Strait. Then you shout 'Honey, I'm home' and the two of you share a quad decaff espresso hot chai with micro-foamed almond milk while admiring late Pininfarina.

So, if public car showrooms are disappearing, then private car showrooms may be increasing in number. At the Sunny Isles Beach tower, in Miami, Porsche Design has included a robotic parking system, so your ride is hauled aloft, and you need never, either emotionally or practically, be separated from your car. And this may be just as it should be.

Driving a car is one of the last meaningful analogue experiences available to individuals: proceeding down the road by means of a more-or-less controlled sequence of explosions within a heavy metal object. Everything else is being sucked down electronic tubes, digitalised, made virtual and fed in big, ugly data sets to the horrible and voracious Mark Zuckerberg.

And when driving is itself outlawed, or becomes as redundant as a pony and trap, or a department store, the last analogue experience will be to admire the subtle aesthetics of an interesting car that is stationary: the evocative stance, fine details, sensuous radii, beautiful proportions, admirable finish. And you may dream elegiac dreams of escape as its designer once did.

But you will need a showroom of your own to do so.

SIMPLE AND COMPLICATED

What is style? According to the poet and artist Jean Cocteau (always very sound on these matters) it is 'for most people … a complicated way of saying very simple things. To us, it's a very simple way of saying complicated things'. And that is an appropriately neat description of the extremely neat Audi TT. For us, it is a simple proposition, but one with many layers of meaning.

Some cars transcend their original humble condition. They actually become more than a sum of their simple parts, acquiring en route the courage of their own restrictions. The Audi TT is exactly that. It could have been merely an aggregate of upcycled Golf componentry. With great ingenuity, that's how Volkswagen created the Skoda Octavia, over which we will solemnly draw a veil.

In fact, purists originally complained that the TT was merely a Golf in high concept drag, but this was neither more nor less than a half-truth. In 1948 car spotters could have made the case that the Gmünd Porsche 356 was a KdF-Wagen in drag as well.

True, the weight distribution inherited from the Golf led to compromised handling which was only partially alleviated by quattro four-wheel drive. But enhanced traction led to some nasty moments on the Autobahn for the over-ambitious entering bends at injudicious speeds: even Audi quattros must obey the laws of physics. And those laws led to accidents. *Alles Mund und keine Hose* they said. All mouth and no trousers. Embarrassingly, there was a recall and that fanatically neat shape was compromised by a mandatory retrofitted rear spoiler.

And then the TT evolved in that typically systematic German way with ever more commitment to ever more technology. In the current, third-generation TT, the dynamic faults have been rectified. So much so that the RS version is at least a match for an equivalently priced Porsche. But the third generation has lost the astonishing, revelatory aesthetic purity of the original.

J Mays, whose involvement in the design we will consider in a moment, said 'I respect it, but I don't love it'. It is the first-generation TT that's the purest expression of the idea: to my eye, the highest expression of the car designer's art. A masterpiece of Cocteau's style.

It was going to be called 'Edelweiss', an Alpine flower that inspired a nauseatingly cloying song in Rodgers and Hammerstein's 1959 *The Sound of Music*. But someone in the Comms Department panicked, pressed an alarm button, and TT was chosen instead. No one seems quite sure, but the reference seems to be to the Tourist Trophy races. And there is certainly family history here: in the sixties, Audi's relation NSU used 'TT' on fast versions of its Prinz. And an NSU motorbike had first won on the Isle of Man circuit in 1911.

April 2019 – the year after the TT's twentieth birthday – was the centenary of the Bauhaus, the most influential modern art school. And there is a connection here. 'Art and Technology: a new unity!' the Bauhaus manifesto declared, although in practice their workshops achieved not a lot more than a useless, geometrical teapot. But the Bauhaus teaching method was radical and its influence profound.

The famous tubular steel 'Bauhaus' chairs by Marcel Breuer were inspired by a bicycle's handlebars: the school encouraged veneration of the machine and of technology. And Bauhaus teachers were inspired by ideas about *Hauptformen* going back to nineteenth-century pedagogy: a belief, almost religious in character, that every object has a significant form.

The last Bauhaus director was Mies van der Rohe, who tried to sell the school's Modernism to Hitler as the authentic German style: mechanical, geometrical, formal, undecorated. But the Führer (who thought 'brown is a very German colour' and liked cuckoo clocks) was having none of it, so in 1933 Mies took off with his colleagues for the United States. Here, he built the Illinois Institute of Technology, in Chicago, and the Seagram Building, on Manhattan's Park Avenue. These were the greatest expression of the Bauhaus aesthetic. At least until the 1998 Audi TT.

Success has many fathers, but failure is a bastard. And the successful TT has in the past twenty years been variously attributed to more than one designer. To achieve some clarity in the confusion, I asked car culture polymath Jürgen Lewandowski to give me a definitive ruling on the car's authorship.

Jürgen told me: 'I have asked Stefan Sielaff [now at Bentley] who was working twenty years ago at the Audi Design Center. And Stefan explained: when the TT was a concept and a show car the head of design was J Mays. The exterior was done by Freeman Thomas, the interior was done by Romulus Rost [also now at Bentley]. When the decision was made that the TT would go into production, the project was handed over to Peter Schreyer [now at Kia]. He is responsible for the TT Coupé and Roadster as we know it.'

Freeman Thomas had spent four years at Porsche before joining J Mays in Simi Valley, deeply inhaling Californian air and perhaps other vapours too, where the two of them created the 1994 Volkswagen Concept 1. This anticipated the TT concept with its pure geometry, confident, bold clean surfaces and the number of resting ghosts it awoke. With great subtlety, Concept 1 evoked the original Beetle without, if we are serious, actually looking anything like it. Similarly, the TT is a clever and haunting reminiscence of the 1936 Auto Union Type C while, objectively, having nothing in common besides a badge comprised of four perfect circles.

I suggested that we photograph the TT at the Alton Estate in Roehampton, London County Council's postwar homage to Le Corbusier. We did this because the TT is such an emphatically architectural proposition. At least, that is, if you find 'architecture' and 'Modernism' to be synonymous. Which is to say: clarity, structural logic, that same refusal of decoration, an affection for austere effects, a narrative that speaks of purpose.

Le Corbusier is the *monstre sacré* of Modernism, although — being a megalomaniac rather than a collectivist — he had only tangential relations with the Bauhaus. He never actually taught at the school and, in that maddening French way, dismissed it as a frivolous '*école d'art décoratif*', its weedy decorators lacking his inflexibly stern purpose. Corb, however, shared the Bauhaus's reverence for machinery. A house, he once declared, is simply a machine for living in.

And he was devoted to cars: handsome automobiles, mostly Voisins, were always artfully placed in the publicity photographs of his buildings. He was photographed on the *pista* above Fiat's Lingotto factory. His Voiture Minimum project (which he boastfully claimed anticipated the 2CV) was, like the TT, a composition of uncompromisingly pure geometrical shapes.

Step into a 1998 TT and it feels vintage. It is tiny and cramped, especially laterally, and the driving position feels hunched and not at all contemporary. And the austerity of the interior astonishes: there are very few controls, reminding us, perhaps, of how wilfully complicated car interiors have become since 1998.

The restraint is artful: again, knobs and buttons and binnacles describe geometry and through half-closed eyes it is easy to imagine the grid upon which it was all designed. There is no fear of leaving a surface undecorated. *Weniger aber besser*. Less but better! It is impossible not to think the designer(s) thoroughly enjoyed themselves here: it is indulgent but disciplined too. If they were asked to back off from an extreme vision, it is not obvious.

The same indulgent disciplines apply to the exterior. The wheel arches are bold semicircles. There is no reason for this except that it looks marvellous. The reversing light, a modest circle, is beautifully integrated into the larger lens. And, like the original 2CV, this light is asymmetrically deployed, with emphatic understatement. Cut lines are a lesson in formal composition. You imagine early proposals might have been composed not with clay, but with Froebel Blocks. The sole concession to decoration is an ever-so-slightly camp filler cap, apparently sourced from a military plane.

I remember introducing the flamboyant socialite and interior designer Nicky Haslam to an early TT. He marvelled at the gaiter around the gear lever, concentric circles of neatly pressed rubber. He ran his fingers admiringly along the rear haunches, commenting on their purity. You cannot call this form-follows-function because that has always been a bit of a nonsense. Form, if it follows anything, follows fiction. And the fiction here concerns a design that describes the romance of driving.

One of the most significant TT owners was the architect, Philip Johnson, Mies van der Rohe's assistant on the design of the Seagram Building. It was Johnson who curated the influential 1932 exhibition at New York's Museum of Modern Art which gave us the term 'International Style' to describe the architecture of Le Corbusier and Gropius. And in his pioneering MoMA car design exhibition, of 1951, Johnson gave us the term 'rolling sculpture'. He used his Audi TT to commute between his Seagram office and his famous Glass House in Connecticut. Perhaps he meant 'rolling architecture'.

DRIVING AND LOOKING

I love that formulation used in US newspapers allowing expert, but indiscreet, comment to continue under the protection of anonymity: 'a person with knowledge of the situation'. Anyway, a friend with knowledge of the situation tells me that the Lamborghini Miura is, in fact, a dog to drive. Heavy, cramped, disagreeably noisy, hot, unresponsive with terrible visibility, secondary ergonomics and quality control that would shame a Moldovan bullock cart. Yet who would deny that the Lamborghini is a sensationally beautiful car?

With cars, there is a most dramatic schism between appearance and reality. If ever you needed to demonstrate the philosophical flaws in the old form-follows-function argument, you could do no better than study the glittering classics of automobile history. Superior function does not necessarily produce superior form. And nor does the presence of superior form indicate an underlying product of any merit. If those last two sentences were not true, we would all be quasi-erotically palpitating about the Soviet-era GAZ-46 MAV amphibious jeep (accessorised with a shovel).

In fact, my own experience suggests no direct connection between great beauty and great performance. The E-type has a slothful throttle, vague steering, uninterested brakes and pitches uncomfortably like a little sailboat in a big headwind. And I am referring to the balletic early six-cylinder cars; later E-types with twelve-cylinder engines drove with no more delicacy and flair than a heavily laden furniture van on deflating tyres.

One afternoon in rural Illinois, I was given the keys to a '55 Thunderbird. Here was the open road and a licence to perform the folkloric psychodrama my book-fuelled imagination had for so long demanded. This heroic Ford was the first mass-produced car designed for pleasure alone. Alas, on Jack Kerouac's road, a '55 T'bird has a woebegone emphysemic iron lump of an engine, a bedridden two-speed gearbox and narcoleptic steering with neither any sense of mechanical connection nor any apparent ability to change the vectors of even a slow-moving vehicle. And yet …

It gets worse. I have always thought the Scaglione-bodied Alfa Romeo Tipo 33 Stradale is, perhaps, the most beautiful car ever made (never mind its cross-threaded screws, pigeon-shit welding and terrible turning circle). Certainly, curves never got any better. And then there is something about its size: it looks both gorgeous and vulnerably delicate at the same time, inspiring a response that can only be described in gross sexual metaphors.

The Stradale's wheels fill the arches satisfyingly and are pushed to the edges of the car's width, creating, if we are talking gross sexual metaphors, a memorable stance. The ratio between wheel height and total height, a metric which designers are inclined to fuss about, is unquestionably correct. There is no slack in the aesthetic composition, just tight'n'toned automobile skin with salacious orifices. Then (unfortunately) you must try to get in. And if you manage the contortionist protocols of access, you will still find the controls all but impossible. If you drive it, you will very soon be turned into a pool of hot human fat. How can the beautiful be so shockingly disconnected from the useful?

The evidence of a disconnect between beauty and utility is profound and continuing. So much so that I have a contrarian inclination to suggest that great beauty may only ever actually arise out of circumstances which never had anything to do with common sense in the first place. Look at Ralph Lauren's oneiric car collection, which was shown in the intellectually lofty galleries of Paris's Arts Décoratifs. A 1931 Alfa Romeo 2300 8C Monza, a 1950 Jaguar XK120 and a 1958 Ferrari 250 Testa Rossa were included. Few would dispute that automobile beauty has any better ambassadors.

One of the organisers of the Paris show asked Ralph Lauren what brings more pleasure, driving or looking? Lauren replied: 'Neither one. You can't divide these things. The beauty of a car for me is a total experience'. I am reluctant to disagree with Ralph Lauren. Indeed, I am wearing a pair of Ralph Lauren trousers as I write, but I wonder if his assertion is really true.

In any objective sense, surely the beauty of a car is divorced from its behaviour. Yes, you can derive pleasure from surviving the neo-lethal mechanical wrestling match that is driving a '31 Alfa, but what's that got to do with beauty? On the other hand, there has never been a more satisfying driving experience than a Golf Mk7. My bet is that Ralph's buckskin chaps will turn to kryptonite before a Golf joins his collection. And yet …

The question is ultimately a big one. What do we mean by a beautiful car? Satisfaction may certainly be had from the act of driving, but beauty is more elusive and, I think, independent of mechanics. Beauty is agonisingly indefinable. I refer to Albert Camus again: '[Beauty] *drives* us to despair, offering for a minute the glimpse of an eternity that we should like to stretch out over the whole of time.'

Isn't that what we sense when we declare a car to be beautiful? Alas, we cannot ask Monsieur Camus. He, of course, was killed when his publisher's Facel Vega HK500 struck a tree, south of Paris. It was the most beautiful French car of its day.

FUNCTIONALISM

'If it doesn't work, we chromium-plate it.' Colin Chapman's legendary – possibly even mythological – words nicely express the perpetual contest between artistic expression and functional performance, which affects all aspects of design.

I was once the judge of the RIBA Awards and a proud client showed me around his shiny new building. I pointed to a certain feature and asked what it might be. 'Oh. That doesn't do anything', he said. 'It's just part of the design.'

There is nothing new here. Classical architecture has a repertoire of details derived from the construction realities of prehistoric wooden buildings now preserved in marble. The metopes and triglyphs on a temple facade don't do anything: they are just visual memories of beam ends and wooden pins, which have a pleasant decorative effect.

Of course, you might expect the functional imperative to be more compelling in the design of sports and racing cars than in the Parthenon. But it is not. The functional imperative is in fact a bit of a fib. As we know, form follows fiction more attentively than it follows the laws of physics. Or put it this way: in the human heart, artistic effect almost always takes primacy over technical logic.

This applies even in aerospace. While an aircraft has to conform to strict laws about drag and lift, the envelope of expression available to airframe designers is actually rather large. Otherwise all planes would look exactly the same. I once met an American aerospace specialist who insisted there was no doubt at all that the McDonnell Douglas F4 was deliberately invested with the semantics of terror.

What I am saying is that 'styling', that term used derisively to suggest the most frivolous and vulgar aspects of design, is ever present. And equally, what else I am saying is the nostrum that strict adherence to 'function' produces beautiful results is one of the great delusions of the modern age.

Sometimes it does and sometimes it does not. As I once observed, the Doxford scavenge pump works very well indeed, but is a thing of beauty only to those of very highly specialised tastes.

A part of the myth about the gorgeous Ferrari 250 GTO is that engineer Giotto Bizzarrini took a model into the *galleria aerodinamica* of the University of Pisa and the result demonstrated the beneficial effects of research in wind tunnels. It was a scientific design! That's only a half-truth. At least as much of the car's expressive power comes from the sculptor's personality of Sergio

Scaglietti, who hand hammered every single panel in pursuit of an artistic ideal. And was it not Ferrari who whimsically painted the cam boxes of a car bright red? There was no functional reason. It just looked wonderful. And anyone who has seen an original Testa Rossa agrees.

By way of contrast, consider the Chevrolet Corvette. The original '53 car was previewed at New York's Waldorf Astoria Hotel, where Cole Porter once sat at the piano in the Peacock Alley lounge and warbled his beautiful, melancholy songs. That gives you some idea of the market General Motors had in mind: it was the cocktail set, not a Sebring pit crew.

But then a Russian engineer, called Zora Arkus-Duntov, joined GM's Chevrolet Division. Arkus-Duntov was a fabulous exotic; married to a nude dancer from the Folies Bergère, he was also a successful competitor at Le Mans. Still, it was his mission to inject some race-bred technical credibility into a cocktail crowd that would have been very happy left with the Corvette's original two-speed auto and lazy, leaden underpowered pushrod six … and a very dry martini. To this end, he wrote a paper called 'Thoughts Pertaining to Youth, Hot-Rodders, and Chevrolet'.

Through persistent lobbying and the waving of this very paper before dull-eyed, martini-injected management, he was able to introduce some technical refinements to the Corvette (including a version of independent suspension and fuel injection). But, ten years on, a collision occurred between race-bred functionalism and sales-led design. For the '63 model year, GM's styling boss Bill Mitchell insisted on a split rear screen for the Corvette, vaguely reminiscent of a Bugatti Atlantic.

The Russian went apeski. As a credible racing driver, Arkus-Duntov knew the real value of proper visibility and decried the stylist's wilful split screen as a travesty of functionalism. After a year's lobbying, the split screen disappeared to be replaced by more functional, but more banal, wrap-around glass.

A great victory of technical purism? Maybe, but the most desirable Corvette is the '63 split-screen car. The unfunctional one created by the stylist, not the version demanded by the principled engineer. Like all works of art, cars do not have to be rational. On the contrary, like all works of art, the most wonderful cars are inspired by quixotic temperaments, momentary whims. And a little magic.

MINI

Henry Ford's 'gasoline buggy' was the most influential car ever, mobilising first America, then the world. Twenty-two million of them changed popular perceptions and possibilities. The veteran *New Yorker* writer, EB White, said 'My own vision of the land … was shaped, more than by any other instrument, by a Model T Ford'.

But almost as influential and much more ingenious was the Mini, which was launched in 1959. Just over five million had been made when production ended in 2000. It shaped not so much a vision of the land, as of cities.

The Mini's raw seams and exposed hinges – honest, unpretentious, unashamed – were an expression in metal of the social liberations of the decade to come. A rough weld was industrial sex and drugs and rock'n'roll. The Mini was not so much an 'instrument' defined by its historical moment, as a car whose various meanings made it a manifesto of the sixties. As the mathematician, Alfred North Whitehead, observed: 'The major advances in civilisation are processes that all but wreck the society in which they occur'.

Because the Mini was so technically and artistically original, it was not classifiable, socially speaking. Hitherto, a small car had been a poor person's car. But the universal appeal of the tiny Mini created a unique social democracy. You could say that it was the very first car that made functional design a positive consumer attribute. Its minimalism was as chic as Mary Quant's mini skirt.

At one moment or other, Minis were very publicly owned by the comedian Peter Sellers, the movie star Steve McQueen, the model Twiggy and Beatle George Harrison, who commissioned a psychedelic-era Magical Mystery Tour paint job for his Cooper to match his colleague John's drug-crazed and fizzing yellow Rolls-Royce Phantom. Enzo Ferrari, no less, owned three S-versions of the Mini Cooper. And the 1969 movie, *The Italian Job*, in which three Minis raced through the piazzas and sewers of Turin, conferred on the car the equivalent of a Papal blessing.

Meanwhile, Ferrari, no Anglophile, had also declared the Mini's contemporary Jaguar E-type to be 'the most beautiful car in the world'. And this was true. On Chelsea's King's Road in the Swinging Sixties, a laboratory of taste where Mary Quant sold her audacious frocks in the Bazaar boutique and Terence Conran peddled his chicken bricks and duvets from Habitat, only two cars mattered: the charmingly utilitarian Mini or the lascivious Jaguar.

The transverse engine with its gearbox in the sump, front-wheel drive, tiny wheels at each corner and sophisticated suspension were technically remarkable, but the true astonishment of the Mini lay on the borders between aesthetics and ergonomics.

There has never been and never will be a car with better packaging: its ratio of internal to external space trembles on the edge of the abyss of magic. And original details emphasised this unique quality: storage bins in the doors and beneath the rear seat, the disappearance of 'the dashboard', sliding windows and string pulls instead of door handles.

And disentangling the authors of the Mini explains something of the democracy of talent which created it. Those charming and cute curves derived from the Austin A30 of Ricardo Burzi, an exotic Argentinian-Italian, ex-Lancia, who found himself in Longbridge after publishing defamatory cartoons of Mussolini. (One of Burzi's Birmingham apprentices was David Bache, who would later design Britain's second-greatest car: the 1970 Range Rover.)

The Mini concept was Alec Issigonis's. He was born in Smyrna into a family of railway engineers. Issigonis, also responsible for the Morris Minor (a car he actually preferred to the Mini), had strict rules: speaking of the Mini's unusual driving position, he insisted people should be uncomfortable since it assisted concentration.

And the ingenious rubber suspension was by Alex Moulton, who lived in a magnificent Jacobean house that was the model for Lutyens' British Pavilion at the 1900 Paris exhibition. Moulton's clever small-wheeled bicycle was another symbol of the sixties. Twiggy used one alongside her Mini. Like Issigonis, Moulton 'never married' and sometimes people talked.

While coarse and uncomfortable, the Mini was in so many ways nearly perfect that no substantial changes were made during its life. Excepting, that is, the 1976 'Clubman' with a meretricious nose-job by Roy Haynes, designer of the Ford Cortina Mark II.

This additional, useless weight was a betrayal of Issigonis's strict, functionalist principles. And it was in a Mini Clubman, leaving a party of Rod Stewart's in Berkeley Square, that Marc Bolan was killed on Gipsy Lane in Barnes one night in September 1977. This, symbolically, was not just the end of Bolan's band T-Rex, but, if we are honest, of the British motor industry itself. The Volkswagen Golf, Renault 5 and Fiat 127 were running away with Issigonis's idea.

True, Harris Mann was working in the seventies, creating designs of great originality – the Austin Princess and Triumph TR7, for example – which the wretched British Leyland, as it had become, was pitifully incapable of realising.

And this is the single source of regret in the whole touching and moving Mini story. What if its manufacturers had behaved with the diligence of, say, Porsche? What if they had patiently evolved details, improved manufacturing techniques, sought better materials and superior suppliers? Would not a well-built and more refined Mini be perhaps the greatest car ever?

But Mini management was inept, short-sighted and too incompetent to make investment-grade decisions. It was unworthy to have inherited an industrial masterpiece as great as this wonderful car. Indeed, it was not until 1976 that they realised they had been manufacturing Minis at a loss. (But Ford, a culture of fixed-cost accountants, had realised this long before. Stripping down a Mini circa 1960, they determined that something this complex could not be sold at a reasonable price. So, they committed to building the crude and chromed Cortina instead.)

I doubt Issigonis or Moulton read Gilbert Simondon, when he was a cult philosopher in the fifties, but Simondon had the clearest understanding of how design works. His book *Du mode d'existence des objets techniques* (1958) can be baffling, but there are passages of hallucinatory clarity. Forgive me if I quote this extract again:

'Artifacts evolve using themselves as the point of departure: they contain the condition for their own development. The structure of the object moves to match the future condition in which it will be employed.'

We are currently into the third generation of the BMW Mini, an evolved artefact if ever there was one. This mutant came about because, by a great curiosity, the Chairman of BMW at the time the company bought the remains of the Rover Group was a cousin of Alec Issigonis.

Today's BMW Minis may be superb examples of sophisticated product design, even if they lack the technical originality of their inspiration. But with their size and weight, they are a travesty of Issigonis's thinking.

Sixty years on, in matters of clarity, purpose and effectiveness, the Mini has never been equalled. Less, as designers always like to say, is more.

BBQ AND WIFE SWAPPING

There's a marvellous saying in Zen that 'whatever is true, the opposite is truer'. You can apply that principle to the question of Japanese sports cars. The question being: are there any great ones?

Japanese culture is stiff with concepts of the superiority of collaborative endeavour over individual expression. They have an expression, *nemawashi*, which translates as 'root-binding', but actually means collective responsibility. Then there is *jishu-kisei* for self-restraint. Hence a public fast train, the glorious Shinkansen, is preferred over a personal idiosyncratic sports car. Moreover, Japan's 60km/h speed limit is among the most stringent in the world.

And yet there is a Japanese sentiment which finds its best expression in sports cars, often of very unusual character. The 1959 Datsun SP211 was based on the Bluebird saloon and called Fairlady, inspired, in that amusing Japanese way, by the company president's 1958 visit to Broadway to see the Alan Jay Lerner and Frederick Loewe musical that was, itself, based on Shaw's *Pygmalion*. Thus, the layered, yet revealing, meanings of automobile nomenclature.

Pygmalion was a play about a hopeless strumpet being civilised through ambition and elocution. The Fairlady followed a similar path of improvement, evolving into a pleasing MGB-like proposition. Although by the time it reached its final 1968 edition what became the Datsun 2000 Sports Roadster had 135bhp, effortlessly shading the English car in every aspect of performance and quality.

Then there was the exquisite 1963 Honda S500 and the impressive rotary Mazda Cosmo of the following year. In 1965 Toyota showed its sensational 2000GT. Clearly inspired by the Jaguar E-type, Toyota refused to attribute its design to any individual until Paolo Tumminelli identified Satoru Nozaki in his fascinating 2014 book *Car Design Asia: Myths, Brands, People*.

Forgotten now is the elegant 1966 Isuzu 117 Coupé, drawn by Giugiaro when he was still at Ghia and at least as fine as the same designer's contemporary Gordon-Keeble. In fact, it's impossible not to believe they used the same drawings twice. But the greatest Japanese sports car of them all was the new 1969 Datsun Fairlady. This we know as the 240Z.

Like all great products, creation myths surround its origins and evolution. But these creation myths, the idea of 'authorship', were a necessary part of the progress and acceptance of Japanese design in the West. Against all the principles of *nemawashi* and *jishu-kisei*, the 240Z has always been recognised as the inspiration of Yutaka Katayama, described in his *New York Times* obituary as an 'ebullient, adventurous man'. Mr K, as he became known, was unlike his timorous and anonymous corporate colleagues. He was one hundred and five when he died, in 2015.

Katayama had been a successful rally driver and became the first president of the Sports Car Club of Japan, an imitator of the American SCCA whose races at Laguna Seca and Bridgehampton offered a theatre for the English sports car. And it was the leading role of MG, Triumph and Austin-Healey that Katayama set to modify with his new Datsun coupé. His story is told in David Halberstam's 1986 book, *The Reckoning*, a study as much about the collapse of the US auto industry as the rise of Japan's.

Because of his extrovert personality, Katayama had been banished from Japan to California, a sort of gulag as seen from Tokyo. As the first president of what became Nissan Motor Corporation USA, Katayama faced derision, cultural obstacles and profound market apathy in America, but under his influence, by 1969 the neat little Datsun 510 saloon was selling sixty thousand units a year. This growing success gave him the prestige and pistonnage to talk his own project into being back in Tokyo.

The precise origins of the 240Z may never perhaps be disinterred from the archives, but it seems to have been based on an early-sixties project called A550X, a joint venture with Yamaha. Albrecht Goertz, a designer who had learnt the craft of self-promotion in the United States from his mentor, the sleek Raymond Loewy, was hired as a consultant.

Hitherto, Goertz had worked on Loewy's Studebakers and the BMW 507 which, since Ferry Porsche was impressed, led to some early styling proposals for the 911. Goertz it was who introduced the Japanese to the use of American-style full-size clay models in the design process, so has a big claim to having begun graphically biased Japan's adventure into Western sculptural 3D.

But the A550X stalled and Yamaha took its engine technology and the rest of the project to Toyota, where it soon appeared in the Toyota 2000GT which Yamaha eventually built in its Hamamatsu factory. Goertz, however, stayed on with Datsun, collaborating with in-house designer Kazuo Kimura on the beautiful Silvia Coupe. But when it was presented at the New York Auto Show of 1965, American critics found the Silvia too cramped and underpowered. This seems to have been the imperative Katayama needed to talk the 240Z into being.

This he did by encouraging another in-house Datsun designer, Yoshihiko Matsuo, who ran Styling Studio No 4, to rage against the conservatives at Nissan who had abandoned A550X, and design a brave, new car. But Goertz stayed long enough to have had his name associated with the 240Z.

Persistent claims by the argumentative Goertz were grudgingly and partially acknowledged by the company in 1980, although Yoshihiko Matsuo and Katayama published a more official list of those involved in their 1999 book, *Fairlady Z Story*. It reads like a musical's cast: Teiichi Hara, Kazumi Yotsurnoto, Akio Yoshida, Sue Chiba, Eiichi Oiwa, Kiichi Nishikawa, Hidemi Kamahara and Tsuneo Benitani. Car design is, indeed, a collaborative venture. And perhaps not one that gives due credit to its heroes.

There is more certain ground to discuss Mr K's concept. He wanted a coupé, not a roadster. This was pragmatic: impending US legislation would soon outlaw convertibles. He liked butch numbers as model designations, not effete names. The 'Z' simply connoted a Jetsons-era modernism. It is said that early designs resembled Giugiaro's Ghibli, but the car that went on sale in the United States on 22 October 1969 had a style all of its own. With its 2393cc 151bhp L-series six cylinder (an engine most honourably inspired by the Mercedes-Benz 180 unit), it easily outperformed English rivals and annihilated the hegemony of MG, Triumph and Austin-Healey.

Katayama said at the New York launch: 'The 240Z represents the imaginative spirit of Nissan and was designed to please a demanding taste that is strictly American … We have studied the memorable artistry of European coachmakers and engine builders and combined our knowledge with the Japanese craftsman.' The car cost a very modest $3,526 and, while some critics found its finish and behaviour a little crude, it soon dominated its class in the symbolically important SCCA races.

Visually, the 240Z is an exceptionally distinctive car. With its long bonnet and emphatically rearwards cabin, the general arrangement is based on the E-type, while its scalloped headlights were inspired by Ferrari, but the whole is unique. It is small but imposing. Aggressive, yet elegant (although most of the original 240Zs had crude *pachinko*-style wheel trims, not proper alloys). Yet it does not look nearly half a century old. But get into a 240Z today and it seems very narrow and feels slight, as well as a bit upright. Doors are insubstantial and strangely thin. The structure predates the computer-aided modelling which, inspired by safety legislation, has given impressive psychological bulk to even the most modest contemporary cars.

The 240Z's glazing bars seem fragile. There are sharp edges and you wince to think of its integrity during an impact. Indeed, a stabiliser bar across the rear hatch opening suggests that body flexing was a problem. The hatch itself closes with a shuddering undamped clang, not a modern moderated thwump. Start the engine and there is a fine induction roar. Press the throttle and there is a lot of noise, but not a lot of progress.

Steering is precise, visibility good. I am not certain I felt that sense of euphoria Katayama described when he said the 240Z gave access to that mystical man-and-machine harmony, but it was certainly amusing to drive. It feels vintage. Sue Chiba's interior with its hard plastics and irrational scattering of tumbler switches and sliders seems Cold War. The 240Z was the first modern Japanese sports car … and also, globally speaking, one of the last old ones.

To my eye, the 240Z cannot be separated from the seventies and its strange visual culture, still influenced by voyeuristic television serials which were themselves located in a more distant, romantic age of one-dimensional heroes and villains following linear plots. It was coeval with the rise of disco and reggae, the avocado-coloured bathroom 'suite', Italian furniture in tangerine plastic, and the decade when Habitat (whose signature colour was a violent green) was the dominant high-street tastemaker with its knock-off bean bags and inimitable chicken bricks. Thus, it represented a gorgeous, remote dreamworld of innocently sexy unwired and sensuously innocent consumerism.

For this reason, we drove the 240Z to the extraordinary Edgcumbe Park estate, in Crowthorne, near Bracknell, on land that once belonged to Windsor Great Park. Here, as my fantasies enlarged, was where you could reliably enjoy barbecues and wife-swapping after a thrilling blast up the dual carriageway from Maidenhead in the Z-car. More prosaically, Edgcumbe was a high-minded garden suburb created by an enlightened developer called Athelstan Whaley who had been influenced by Scandinavian domesticity and the ranch-style houses of California. As Katayama said, the 240Z package was addressed to America. As was most advanced design at the time.

Exactly contemporary with the 240Z and its Fairlady predecessor, Edgcumbe Park was begun in 1958 and completed in 1970. The ambitious brochure – more, really, a manifesto – was revealing: '*The* place to live West of London', it said. 'Every house, every site and winding cul-de-sac is imaginatively planned by our Architectural Staff [note CAPITALS] preserving the Oaks and the Mountain Ash, the Scots Pine and Sycamore, ensuring good orientation and pleasant views'. And, if you could afford it, you would have a Charles Eames chair and ottoman next to your heated serving trolley with its taramasalata and beef olives around which your female guests would gather wearing billowing cheesecloth dresses and agreeable pouts.

It was in a house on this estate that François Truffaut shot, with Julie Christie, his film of Ray Bradbury's *Fahrenheit 451*, a novel set fifty years in the future. The estate has been described as 'the future that time forgot'. And now, as we remember it, how distant that age of thigh boots and hot pants seems. Delicious to imagine the felicity of driving a fast and reliable 240Z home, parking it in the drive of your California-style ranch home in Berkshire, then sipping a Campari and soda before you enjoyed a casserole served in bright orange (Raymond Loewy designed) Le Creuset oven-to-tableware.

Over half a million 240Zs were manufactured and its success lent Datsun an aura of prestige that could not have been achieved by the contemporary front-wheel-drive Cherry. As the *New York Times* noted, in 2008, it changed 'the auto industry's perception of Japanese cars'. Katayama-san retired in 1977 when Japan was still a pompous, conservative and hieratic nation.

While America acknowledged his achievement with the 240Z, at home his high profile was interpreted as vainglory and Katayama was not feted in retirement. But with the increasing scholarly interest in the history of car design, Katayama began to emerge as a significant figure and by 1997 Nissan was running television ads featuring the ebullient Mr K, father of the Z-car.

The 240Z is one of the great Japanese cars. In fact, one of the greatest cars of them all. Now, consider again that original Zen proposition and just take pleasure in the ability of cars, good and bad, to evoke a powerful and romantic range of ideas. Great cars take your imagination, as well as your body, on fascinating journeys to remote worlds. Even as far as Crowthorne.

UGLY CARS

In the final act of the Cold War, when Berlin still smelt of Russian petrol and Russian women still smelt of cabbage, Californian artist Phil Garner had a project called 'Send a Gremlin to the Kremlin'. The Gremlin referred to was one of the last cars produced by a demoralised and moribund AMC. A cut'n'shut Hornet sub-compact, it was agreed, by experts and amateurs everywhere and without a single voice of dissent, to be one of the most shockingly ugly and maladroit cars ever seen. It was magnificent in its defiance of taste and intelligence. Impressive, in a perverse way. Garner's idea was that the CIA should use the hideously stunted Gremlin to give the Soviets damaging misinformation about the competence of American industry.

Cars are meant to be beautiful, to excite desire in customers and admiration (usually accompanied by envy) in onlookers. It is no accident that the word 'consume' has connotations that are both retail and erotic. So much of the historic adventure of car design has been involved in chasing down the fugitive idea of beauty, establishing the metrics of refinement, graduating the scale of desire, gratifying the senses.

But there is evidence that the automobile's aesthetic adventure might be over. Or, at least, abandoning that version of it and shifting the paradigm. Some years ago, when the last generation BMW 1 Series was still a rare sight, I was having lunch with the urbane J Mays, then Ford's creative boss. A 1 Series came into view and we stared at its strange sculpted panels, warped proportions and uncompromising toad-like stance. 'Christ! That's a shit ugly car!' J said. Anybody disagree?

The interesting thing here is that automobile ugliness used to be caused by ignorance or incompetence. That Gremlin! Or the get-away-from-me nightmare of the SSangYong Rodius! But BMW was neither ignorant nor incompetent. And nor was there the excuse of budget, although this would be a poor defence: very often exiguous resources stimulate designers into inspired creativity. BMW is instead one of the most impressive industrial undertakings in history, with a dazzling record of authoritative achievement.

Ever since Max Friz's R32 boxer engine reflected and projected Bauhaus principles of art and technology, BMW has got it right. And here was BMW volunteering a design that was difficult to look at. This was automobile ugliness as an act of free will, not as an embarrassing accident. All those serious and crisply dressed PhDs on the BMW board, who lived and worked in Munich, one of Europe's most cultivated and art-drenched cities, stared at the awkward, unlovable 1 Series proposal and solemnly nodded it through.

Because it is more advanced than many other manufacturers, BMW reached the 'beauty crisis' first. For so many years BMW design had followed a path of Bauhaus clean lines, of intellectual refinement, of severe, but elegant, *gute Form*. And it had become boring. You can have too much perfection, so in the early years of the twenty-first century, BMW reversed assumptions about beautiful cars and for the very first time in the history of design, a manufacturer decided no longer to seduce and gratify the customer, but to challenge, confront and, let's be honest, annoy him.

It's what's called destructive creation. You destroy in order to build. People groaned about a sequence of BMWs with slab flanks, irrational wobbly surfaces and an overall lardy grossness, but something very interesting had occurred: car design had changed. Every new car now has something of BMW about it. Since beauty has always been so very difficult to define, so fragile, why not abandon the struggle and make visual noise the new objective? Bauhaus perfectionism was defined by architect Mies van der Rohe when he said, 'I don't want to be interesting. I want to be good'. Designers now reverse that.

Porsche did it too. The first-generation Cayenne was both conceptually ugly (an affront to Dr Porsche's noble conceptions of efficiency), and visually unsettling (an early 911 dropped from a great height on to an innocent garbage truck). It has been category-bustingly popular. Then there is the Panamera, fat, heavy, ham-fisted: as gorgeous as a lacquered elephant seal. They cannot make enough of them. Beauty is commonplace; the new challenge is ugliness. It's fascinating.

I imagine they are discussing this in design schools. If you ask a student to design something ugly, they find it difficult. But there are other challenges with ugliness. If everything were beautiful, nothing would be. Beauty tends to sameness. On the other hand, the variety of ugliness is infinite. As the grizzled old chanteur, Serge Gainsbourg said, 'ugliness is superior to beauty … it lasts longer'. We shall see.

ROLLING SCULPTURE

New York's Museum of Modern Art is one of the West's great institutions. It's one of my favourite museums. But that doesn't mean it's not also very annoying. From its home on West 53rd Street, for more than seventy years it has promulgated the democratic spirit of Modernism. But its founder was billionairess Abby Aldrich Rockefeller, Mrs Standard Oil, a woman not known to stand on the barricades of proletarian revolt. It is still run by an exclusive clique.

Nonetheless, it was MoMA's hugely popular Picasso retrospective, of 1939, that created its own reputation and cemented the artist's. His greatest work of protest, *Guernica*, was, in the terms of his bequest, on display in New York until the Fascists left Spain (even if they remained on Manhattan). The exhibition returned in 1981.

MoMA was the first gallery anywhere to have a serious design collection, suggesting that beautiful kitchen machines have the status of art. Which, in my view, they do. And in 1968 there was an amazing exhibition called *The Machine as Seen at the End of the Mechanical Age*, curated by Pontus Hultén. The catalogue had an astonishing metal jacket and still stands as the best account of how technology and art so fruitfully miscegenated in the twentieth century. It was an epoch-defining publication.

And, earlier, MoMA was the first fine art museum to take cars seriously. In 1951 the exhibition *8 Automobiles* was built on a ramp covered in white pebbles that swooped from the first floor into the garden. The press release, written on a wobbly mechanical typewriter with the odd jumping key, explained that on show would be 'cars selected for their excellence as works of art'.

The curator was Philip Johnson, whose ancestors had helped Peter Stuyvesant lay out the plan for New Amsterdam. Johnson was rich, gay and got ever so slightly too close to the Nazis when visiting Germany on research trips before the War. And he became one of the most influential tastemakers ever. First, he introduced Mies van der Rohe, the last director of the Bauhaus, to America, and in the fifties became his great architectural collaborator and rival. Then in the seventies, Johnson changed his mind and became the champion of Postmodernism. 'Remember, son, I'm just a whore', he once told me.

But since architecture and prostitution are so closely related, it's interesting to see how Johnson applied architectural criteria to his assessment of a car's design value, as seen from 1951. He liked the 1931 Mercedes-Benz SS because it was 'on a heroic scale with each detail appropriately developed for the total effect'. The Jeep was admired as a 'genuine expression of machine art'. Both the Benz and the Jeep made no attempt artistically to integrate the separate elements of wheels and passenger compartment. That was an effect to come later.

Briggs Cunningham lent a razor-edged 1939 Bentley to the exhibition, and there was also an MG TD on show. Johnson admired the little sports car because, while remaining essentially a mechanical proposition, the designers had added interesting visual enhancements. The arabesque of the front wing, for example, or the judicious use of artfully highlighting chrome instead of the 'meaningless decorative strips' usually found on American cars. You have to imagine the MG close to *Guernica*

Then there was the emerging 'single envelope' design of Pininfarina's Cisitalia. This car, bright red, is still on display in MoMA's permanent collection. Or a Talbot with its 'expressive use of streamlining'. America was represented by a 1937 Cord and a 1941 Lincoln Continental, although Earl 'Madman' Muntz's Jet, a Frank Kurtis design recently published on the cover of *Popular Science*, was shown in photographs. So too were 'two rear-engine cars, both designed in Germany by Dr Porsche'. These were the Volkswagen and the Gmünd-era 356. Johnson noted that the latter 'illustrated the extreme development of the seemingly one-piece metal lid'. This they are still doing.

No one has better stated than Philip Johnson the proposition of cars as art: 'Automobiles are hollow, rolling sculpture, and the refinements of their design are fascinating … besides being America's most useful Useful Objects [they] could be a source of visual experience more enjoyable than they now are'.

Johnson's words were in my own mind when, in 1982, I put the very first car on show in the Victoria & Albert Museum. This was a Saab 92. A year later, I showed a Mark I Ford Cortina, the one with the CND rear lights. For the opening of London's Design Museum, I had a modelmaker build a full-size wooden version of Le Corbusier's Voiture Minimum.

And so, I often think, as we near the end of the Age of Combustion, what would *8 Automobiles* comprise today? Could anybody ever agree on precisely what eight cars were true works of art? I don't think so. The great thing about the soon-to-be-past Age of Combustion was its practical and aesthetic certainties. These have disappeared, like exhaust smoke.

CINQUECENTO

Nineteen-fifty-seven was a year of powerful symbols, both large and small. The Soviet Union humiliated the United States by launching the first artificial satellite. And, since this was propaganda as much as rocket science, the technicians at Star City were told to make *Sputnik* look attractive. This they did: the polished blob with trailing antennae remains one of the most evocative designs of the Cold War.

Meanwhile, back in Europe, we, by which I mean 'they', signed the Treaty of Rome. In Turin, Lavazza built the first modern coffee roastery on Corso Novara. It was capable of handling 40,000kg of beans a day and can, legitimately, be said to mark the beginning of the International Cappuccino Era not yet even now near its frothy and caffeinated peak.

And at the Turin Salone that year Alfa Romeo showed Bertone's Giulia Sprint Speciale. What a fabulous car, but on the whole, the little Fiat Nuova Cinquecento, also a product of 1957, has lasted better. Fiat's Dante Giacosa has some claim to have been one of the great geniuses of the automobile business and the little 500 was his consummate masterpiece.

I think it may be a general truth that people with an educated interest in cars may also be inclined towards an Italophilia that's situated somewhere between respectful enthusiasm and demented obsession. They say Italians do not separate life and art. Thus, creating a humble car that is a pleasure to drive and a delight to see is, in its way, similar to treating the making of an ordinary, but perfect, espresso a rite of religious seriousness. Who now remembers that Italian cars were the last to be fitted with cupholders? Why? Because coffee is one thing and driving is another. It would be sacrilegious to muddle the two.

The Nuova Cinquecento was successor to the Antico 500 of twenty years before. And it was a successor in a different way to the Vespa of 1947. The scooter mobilised postwar Italy and, ten years on, it was sure evidence of the success of Italy's *Ricostruzione* that a single cylinder, wobbly two wheels, rasping two-stroke, handlebars and a pillion had been replaced by four wheels, four strokes and (approximately) four seats: the tiny Cinquecento remains one of the best packaged cars of all time, a miracle of spatial management. It was a miracle of economy too: very special care was taken to maximise efficient use of the sheet steel, minimising waste. Giacosa even wrote an article about the clever metal stamping in *Stile Industria*, the Italian design magazine.

And, of course, the result was a car of exceptional 'charm' although, significantly, the Italians have no such word. It was a Nuova Cinquecento that tempted me into my only experiment with classic car ownership. Bought on a whim when I saw it parked on King's Road with a 'For Sale' sign on the windscreen, its idiosyncrasies became endearing. The twin levers on the floor for starter and choke! The workshop manual instruction suggesting the use of tin snips to obviate the inconvenience of a jammed bonnet! The daily losing battle with fast-moving rust! If there was synchromesh, I never found it. Not to mention its absolute refusal to start from hot, necessitating careful planning of travel itineraries. But it was a severe delight to use, requiring the skills of both a trucker and a racing driver since maintaining momentum was necessary if continuing progress was a goal.

I even contemplated trading up to an original Abarth 595, eventually finding a bright blue one and testing it around Clapham Common, pretending the A3 was the Targa Florio. I adored the neat alloy wheels with weird camber at the rear and the propped-open engine hatch, as titillating, as Alberto Moravia, a great novelist of 1957, might have said, as the gap between knickers and stocking-tops (since these were the days before elasticised tights enhanced convenience and reduced glamour).

But there is so much questionable about the provenance of these cars. Not least because Carlo Abarth himself, an associate of Porsche and an employer of Aurelio Lampredi, made so many retrofit components (including his famous exhausts, now collectors' items in their own right) that 595 tributes way outnumber 595 originals. Only the reckless would buy one without background checks and I am not reckless. So, it was good, for the avoidance of doubt, when, soon after the *nuova*-Nuova Cinquecento of 2007, Fiat revived the Abarth brand with a new 595. I have been driving one, and it prompted this meditation. Others can better describe the dynamics of a car which made me smile continuously, but trust me as an absolute authority on Italian sensations: *E la cosa reale.*

The car moved my imagination a few years on from 1957 to 1961. In that year, Italians got the Splügen Bräu pub on Milan's Corso Europa by the Castiglioni brothers: the best ever modern bar. We got Watney's dismaying prefab pubs on council estates. To me, in design terms, cars and architecture are much the same thing: moving environments or static ones.

In whatever vintage, the Cinquecento proves the old Modernist adage that the best is always simple, but simple is not always best. It is a car of low cost, but high culture. And if you want a convincing demonstration of Italian superiority in these matters, after you have Googled Abarth 595, Google 'Brabham Viva'.

SNOBBERY

I have been compiling a list of the greatest ever car designers, excluding
the dross, the momentary celebrities and trying to get down to the absolute
essence. I mean: who are the very few people in the past century who have
given astonishing shape and enduring meaning to the idea of the automobile?

The astonishing thing is, there's not a single Briton on my first draft. This is
extremely strange because, despite the comic and tragic vectors of Britain's
industry, we are still one of only seven or eight countries for whom the car is an
expression of a singular culture. And there are one hundred and ninety-six
countries in the world. For the avoidance of doubt, the other significant car
manufacturers are the US, France, Italy, Germany, Sweden and Japan. We can
argue about Korea.

Let's define terms. I have been worrying about 'design' for so long, it may be
time to give up. Maybe there is no such thing as design, but there is certainly
such a thing as a designer. These are individuals who are part artist and part
technician. While they may understand the machinery, they are not engineers,
but visionaries who see the world in a certain way and are determined to make
meaningful contributions to it.

The shortlist is not difficult. There could be no debate about (Nuccio)
Bertone, (Flaminio) Bertoni and Pininfarina being on it, nor Giugiaro and
Gandini. And I could easily defend a place for Bill Mitchell, Giovanni Michelotti
and Robert Opron. Mitchell's Oldsmobile Toronado was the last great American
car, while Opron's three great Citroëns (the SM, GS and CX) plus the very odd
Renault Fuego are splendid eccentric genius.

On Michelotti, I would confer a special award for bizarre range: from the
Triumph Herald to the Alpine A110 is a long journey indeed. I would include
Patrick Le Quément too: while no great beauties, the cute Twingo and clever
Scenic changed the shape of cars forever. Possibly Virgil Exner: no one who
has seen a 1957 Chrysler 300C will ever forget it. A vision kept fresh by
recurrent nightmares.

The difficulty comes in considering supremely talented individuals whose
output was small or had only slight influence. Gordon Buehrig and his Cord
810 is one example and Sixten Sason's various Saabs are others. So too Ercole
Spada and Piero Castagnero: the Aston Martin DB4 GT Zagato and Lancia
Fulvia Coupé remain exquisite, but were never popular, nor successful (and
those two qualities are attributes of design).

Another methodological problem is those extremely influential individuals who were really more inspirational managers than lonely visionaries. Do we include people, some of whom never actually picked up a pencil, on a list of designers? There are a lot of them: Harley Earl of GM, Wilhelm Hofmeister of BMW, Leonardo Fioravanti and Lorenzo Ramaciotti of Pininfarina and Uwe (Sierra) Bahnsen and Jack (Mustang) Telnack of Ford. I think we do.

So, what happened to the Brits? For some reason I kept on excluding William Lyons. Although everyone rightly adores his great, beautiful Jaguars, I do not think Lyons identified himself as a designer, and self-awareness, not to say rampant ego, is a defining characteristic of this tribe.

Sure, Frank Feeley and Walter Belgrove deserve honourable mentions for the Aston Martin DB1 and Triumph TR2, but these cars are pleasant idiosyncrasies, not industrial events of world-historical importance. And from today, Ian Callum and Gerry McGovern have done excellent work at Jaguar and Land Rover, but the solemn Muse of History is not yet admitting them to the All Time Hall of Fame.

Yet, following a national taste for self-effacement, I wonder if I am being biased. There was, in fact, a moment during recent history when British designers were doing unique work. There was Bill Towns at Aston Martin, David Bache at Rover and Harris Mann at the chaotic amalgam that eventually expired as BL. The great Italians we love because they make us think *la dolce vita*, American cars are rock music and everyone loves rock music, the great German cars were positive expressions of the *Wirtschaftswunder* when everyone got rich, while French cars evoke images of either sumptuous luxury or elegant chic.

But Towns, Bache and Mann had the misfortune to be working at a historical moment no one much liked at the time and no one wants to recall later. Their culture was industrial malaise, cultural stagnation, political impotence and economic calamity. And yet they worked against it with designs of great optimism and originality which, people will eventually see, constitute a unique British contribution to this most evocative subject. The hopelessness makes it the more touching.

Of course, their efforts were compromised by dim management and poor execution, but isn't it only snobbery that prevents us from seeing the Rover-BRM gas turbine racer as the most sensational Le Mans car of its day, the Austin Allegro as an ingenious package well ahead of its time, and the Rover SD1 as a uniquely bold reinvention of the traditional saloon? Yes, it is only snobbery. I am going to revise my list.

THREE EXCEPTIONAL MACHINES

Jaguar XK120, Felixstowe Rally, 1952

The beautiful Jaguar XK120 caused a sensation when
it was launched at the 1948 London Motor Show. And
Jaguar's owner, the highly entrepreneurial William Lyons,
capitalised on that sensation by organising a memorable
publicity stunt the following year. There being no roads
straight enough in England for the high-speed runs
he planned, the Jaguar team went to Belgium. On the
Ostend-Jabbeke highway, the XK120 was timed at
132.6mph. This, when a family car would struggle to
achieve sixty. Jaguar's reputation was made. While an
emphatically British product, the style of the XK120
was inspired by the prewar BMW 328.

Ford Edsel, 1957

In September 1957, fifty years after Henry Ford realised
that the gasoline buggy was a 'universal desire'
(enthroning the ordinary American in the process),
his successors launched the calamitous Edsel. *Time*
magazine described this $250m mistake as 'a taste of
lemon'. A psychologist blamed its commercial failure on
the public's unconscious reckoning that the radiator grille
looked like a terrifying chromed vagina approaching at
60mph. Inside the car, from left-to-right, Henry Ford's
grandsons: William Clay Ford, Benson Ford and Henry
Ford II.

Triumph Herald Convertible, 1959

Harry Webster wanted to vivify the lacklustre Triumph
range and chose Giovanni Michelotti to carry out the
life-saving operation. The late fifties was, of course, the
moment when Italian popular culture first entered British
consciousness. Cliff Richard's film *Espresso Bongo* was
set in a Soho becoming flush with Italian-inspired coffee
bars. Fellini's *La Dolce Vita* was released in 1960, and
soon, men began wearing Italianate 'Slim Jim' ties. And
Turin arrived in Coventry. The Herald's extraordinary
angularity was a result of the manufacturer lacking the
sophisticated pressing tools needed to bring Michelotti's
curvaceous design to life.

ALL CARS LOOK THE SAME

There are some years when more happens than usual. Caused by climacteric oscillations, who can say, but 1848 was, for example, a year of remarkable revolutions in all of Europe. So much happened in 1913 that Florian Illies wrote a whole book about it with lots of Viennese angst, but only after Liliane Brion-Guerry had devoted three entire volumes to the same twelve months with lots of Parisian elan.

Personally, 1957 continues to fascinate. Notable for the last Jaguar D-type win at Le Mans, and the launch of *Sputnik*, it was also the year that Bill Haley and the earthbound Comets landed at Southampton, introducing Britain to rock'n'roll. Plus, the Chevrolet Bel Air, Detroit baroque at its giddy peak. Goodness, what a busy year. Then there was 1967.

One of the most boring tropes is: 'all cars nowadays look the same'. I actually never mind when people are hostile in this way. There's no argument that the automobile offers only limited evidence of man's ingenuity and artistry, while providing irrefutable evidence of his selfishness, brutality, venality, stupidity and general inclination to destroy the planet and its occupants. Still, it's true that all artefacts created in a common historical era exemplify the Zeitgeist, or Spirit of the Age, and why should they not? To be sure, shared circumstances often lead to similar solutions to shared problems. But 1967 confounds that.

It was the year that gave us the Alfa Romeo Montreal: Marcello Gandini asserting himself at Bertone. Claus Luthe's NSU Ro80? Still a technical and aesthetic achievement that astonishes. Dino 206? The most beautiful Ferrari ever. By contrast, the Fiat 125 was an Italian Lotus Cortina, but with much better seats and marginally superior reliability. The quirky Saab 99 was a Mini designed by aerospace engineers. So that was a good thing.

Or the Citroën Dyane, a de-luxe 2CV. One version was called 'Weekend' because, the joke went, it was so slow that it took two whole days to get anywhere. I had one. Beige with brown perforated vinyl trim. It took me all over Europe, and I once mended the throttle linkage with a wire coat-hanger. So that's a V8, a twin-rotor Wankel, a V6, a twin-overhead-cam four, a single-overhead-cam four that would eventually birth the first popular turbo, and an air-cooled flat-twin. And the designers found satisfyingly different artistic expressions for these gloriously different engines. All cars look the same. Are you mad?

It is tempting to think of 1967 as a miracle year for car design. Will such variety ever be repeated? Certainly, a decade on, 1977 had far less to offer. This was, for example, the year of the Daewoo Maepsy (a kimchi-flavoured Opel Kadett) and the Peugeot 305, a car that marks the beginning of that manufacturer's descent from principled ingenuity to anhedonic market-led banality. We will pass over the Triumph TR7, whose strange, scalloped lateral panels made Giugiaro say: 'Dear me, have they done that on the other side as well?'

But there were two cars from 1977 that demonstrated man's perverse genius in finding idiosyncratic answers to questions not many people were actually asking. Each a big, front-engine 2+2 with ample power and luxury appointments, offered by manufacturers with proud reputations for engineering integrity. Winking at each other like Gustav and Hermann in a north-European gay sauna, I give you the Volvo 262C and the Porsche 928. If I ever become disenchanted with looking at cars, I will return to this article to refresh my sense of awe at the superhuman curiosity of these machines.

The only explanation for this most curious Volvo is that, while they were discussing a 'personal' car for the US market, someone slipped magic mushrooms into the lingonberry compote. And, in a bit of a hallucinogenic muddle, someone promptly called Bertone in Turin and asked him to turn the cumbersome six-cylinder 164 into a coupé. The result was one of the most shockingly ill-proportioned, maladroit and odd cars ever made. Arctic snowshoes with Armani details! It was superlatively ugly, thus fascinating.

Only slightly less curious impulses gave rise to the 928. Its designer was Wolfgang Möbius, working under Anatole Lapine, who claimed that his education had been at the University of Hard Knocks. Thus, the company founded by a professor of daunting academic attainment left the design of its most ambitious product to a pugnacious Latvian-American who wanted to make a German Chevrolet Camaro. It's these cultural absurdities that make car design so fascinating.

The Volvo 262C and Porsche 928 have all but disappeared from the roads. Yet I suspect each will be back: the anthropologist's 'myth of the perpetual return' predicts it. I am less confident about 2027's potential for leaving us with future classics. But 1967 and 1977? Look back not in anger, but in wonder.

PROBOX

'Big coffee for luxury fun life!' I have just made that up, but it is typical of Japanese advertising. I adore Japan and have been visiting regularly for thirty years, but the experience is still bemusing. Every attempt by a Japanese to explain anything only leads to further mystification. The language is not so much a barrier as a camouflage disguising a completely different habit of mind.

An educated Japanese might know five thousand of the Chinese-derived Kanji characters. This is why you find such extreme inventiveness in the naming of things: they run out of Chinese. This is how Sony's Akio Morita arrived at 'Walkman'. Mind you, 'Sony' itself was a happy invention wherein earlier Morita had muddled the Latin word for sound with the affectionate term for a male child.

I arrived one summer at the beginning of the rainy season. Already somewhat dazed by a twelve-hour flight (during which I accidentally watched the new Kevin Costner film three times), in the downpour on the flooded airport road I am certain I saw a car called a Toyota Noah. Then my imagination took over. Was that really a Nissan Proboscis? Surely not a Mazda Flatmate? An Isuzu Hangover?

Then there are the categories of cars they make, with products occupying niches we cannot even imagine. A favourite of mine is the Toyota Alphard, a blingingly bejewelled and bordello-plush MPV much favoured as a shuttle by luxury hotels. The performance version is called Vellfire. Nissan's equivalent is the Elgrand.

No wonder the Japanese popularised the term 'hybrid'. If cars were biological material, being on the streets of Tokyo would be like having access to a mad professor's laboratory with the mutants running wild. But remember that while a mutant is certainly an abnormality, it is the product of a deliberate decision, not of error.

So, I am wondering what powerful mutagens were used to create the Toyota Probox. A popular taxi in South America and Myanmar, the Probox (launched in 2002) was designed for Japanese tradesmen. Aesthetically, it is one of the most fascinating vehicles on sale. Not because it is exciting or interesting, but because it is uniquely boring. When the rest of the automotive world is straining to add semantic value to its products, Toyota has found in the Probox a way to signify absolutely nothing.

It is stripped back and featureless. You could not call it minimal because Minimalism in comparison looks like a Baroque fantasy. Merely looking at a Probox drains energy from your system. But there is a perverse sort of genius here: the neutrality achieved by this featureless, motorised and glazed cannister induces in the viewer a condition of Zen-like meditative calm. The art colleges of the world are overpopulated with students anxious to 'do' another hypercar. Extremes are easy to explore because you know where they are: it is the lower-middle ground that is truly uncharted territory.

Historically, the Japanese artistic tradition has reached greater heights in two-dimensional graphics than in three-dimensional sculpture. Japanese cars once reflected this: very good details, but weird stance and angles. Then, about twenty-five years ago, computer modelling allowed Japanese designers to experiment with more expressive solid forms. This they have ever since done with some abandon, yet somehow the Probox avoided the stain of creative input. It is the least impressive car you will ever see. Therefore, one of the most fascinating.

I can imagine now the Probox gear-change, both rubbery and precise in that contrarian way that Zen favours. The engine, subdued, but whining, in its efforts to drag the bulky Probox from plumber's depot to blocked domestic loo. The grey plastic slippery seats that are neither comfortable nor uncomfortable. The tumbler switches for the electric windows, which feel cheap, but work exquisitely. In my imagination, I feel that the Probox is perhaps like Ferry Porsche's definition of the 911: 'driving in its purest form'. Not, of course, an exciting form, but an uncontaminated one, for sure.

Frankly, I am bored with Lamborghinis. Outside my Soho office one day, someone parked an Aventador in a migraine-inducing metallic beetroot colour with, as I recall, black suede upholstery. I didn't give it a second look, but while I was in Japan, I positively sought out the Toyota Probox for aesthetic gratification. There's probably an app that does hypercar shapes, but the Probox is a rarer, finer thing. It's what John Updike called 'giving the mundane its beautiful due'.

Of course, you will never hear or see a Probox in a classic motoring magazine. The only Japanese cars likely to feature there are the Datsun 240Z and the Toyota 2000GT. But each of these was influenced by European models (as well, perhaps, as the swift hand of the mercurial Count Albrecht von Goertz). The Probox is indissolubly Japanese and that gives it great character: uncompromised national identity and absolute authenticity are part of any definition of 'classic'. In comparison, an Aventador looks a frightful muddle.

AIR INTAKES

My subject here is orifices. A car's principal air intake and its close relations, the louvre and the duct, have often been the defining features of a design. But now, just as public smoking legislation has doomed the minor art form of the commercial ashtray to extinction, so new automotive power systems and sophisticated aerodynamics are diminishing the role of the hole.

Still, I have a fanciful idea that you could write a history of car design by studying the shape of air intakes alone. If you want examples of the extraordinary aesthetic subtleties, games of proportion and nuanced semantics that designers so artfully deal with, you could do no better than immerse yourself in the automobile orifice as it evolved from the mid-fifties.

Put it this way, I think most readers would be able to identify '1957 Vanwall' simply from the shape of the Frank Costin-designed air intake. In terms of visual language, the stuff that preoccupies aesthetes, that's an extraordinary testament to the power of design.

I like the orotund magnificence, with echoes of the language of Dante, that Pininfarina uses to define the *presa d'aria* in the valuable little *Lessico della Carrozzeria*: '*Imboccatura attraverso la quale il flusso d'aria viene canalizzato nei punti richiesti per il raffreddamento di parti meccaniche o per raffrescare l'abitaclo*'. Like the most moving parts of Dante, this is best left unmolested by translation, but if you take anything as seriously as Pininfarina takes the air intake, you will be bound to pay close attention to its design. The 275 GTB? Could a hole be more beautiful?

Basically, heat engines and brakes need cooling, and so too do the passengers. Thus, the need for a variety of holes to capture and channel air for this purpose. It is because these holes have some of the characteristics of a mouth and nostrils that the front elevation of a car sometimes resembles a human face, a little bit of theatricality in which the lights play the part of eyes. As Charles Darwin showed in his 1872 study *The Expression of the Emotions in Man and Animals*, the mouth is peculiarly articulate. Just a little flexure of the lips can change the meaning of a facial expression from ecstasy to revulsion via doubt and anxiety to joy or agony.

This human parallel gives the air intake such gestural power. To understand how science must defer to art in making a car beautiful, consider the revolting air intake of the 1952 Cunningham C-4RK. This was the work of the pioneer aerodynamicist Wunibald Kamm and it looks, if we are honest, like a puckered orifice where the sun is not known to shine. Compare it to the 1958 Scarab, once routinely cited as 'the most beautiful racing car ever made', and I see no reason to argue with this assessment. This was the work of Lance Reventlow, not a scientist, but a playboy heir to the Woolworth fortune. The Scarab's major air intake is a delicious slit, a beautiful, crushed ellipse taken to exquisite points at its extremities. Sublime.

The E-type comes rushing to mind. The original car had a gorgeous aperture with sexual suggestions there for all to drool over and boggle at, a major psychological factor, surely, in the car's popular success. The later V12 needed more cooling, thus a bigger intake which compromised the art as much as the heavy engine ruined the handling. A few extra horsepower was bought at a terrible aesthetic cost, as it often is. And now there is the F-type. I know they agonised about it, finessed the radii, pondered the proportions, studied the data, sent the car to clinics, but I am sorry: no. If you want to see how a square intake works artistically, consider the '64 Mustang or the '66 GT40.

The minor intakes were just as important in design terms. At the historical moment when aerodynamics was being promoted from guesswork to technology, cars began to appear with NACA ducts. The acronym belongs to NASA's predecessor, the National Advisory Committee for Aeronautics, whose wonderfully expressive intakes were depressed flush into the bodywork so as not to cause drag. Shaped in plan a little like an architect's ogee arch, they created vortices that caused suction which multiplied the volume of air taken in. The history of the NACA duct begins with the experimental North American YF-93 interceptor, of 1950, and ends thirty-seven years later with the Ferrari F40.

In between, gloriously kitsch fake ducts appeared on various Mustangs and Corvettes, evidence, perhaps, of Detroit's incipient depravity. But, actually, the 1989 Ferrari 348 had a front intake that is entirely bogus. Imagine the Pope in a wig! I like vulgar fakery in America, but not in Italy. That was the moment Ferrari began its aesthetic decline. As I say, you could write an entire history of car design by looking at holes alone.

BRISTOLS

I have been specifying my next car. While still very – some might say 'extremely' – vigorous in the social, professional, athletic, romantic, oenophiliac and gourmet senses, I am, nonetheless, now so (very) old that I can (just about) remember when a heater was an extra. My father once had a Mini as a spare car and I can recall the excitement when he bought an after-market kit that allowed him to flash the headlamps with a wand.

So, this is all a bit exciting. Anyway, my new car is going to have laser-guidance, high-definition forward-looking infra-red (HDFLIR), helmet-mounted cueing systems, ruggedised data links as well as beyond line-of-sight communications and video. It will self-park, self-wash, self-tax, self-refuel when the batteries lose charge and, indeed, self-destruct when the owner acquires twelve points. And in this way we will have arrived at a Jerusalem … of sorts.

My father used to work in the aircraft industry, and childhood trips around the factories he managed are profoundly rooted memories of mine, as is the aroma of hot oil and swarf, since the part of the brain that processes smell is adjacent to the part that files our recollections. He always used to interrogate me about how things were made. Handed a childhood bottle of pop, he would demand my answer about whether the bottle had been cast, moulded, carved, milled, machined, turned, sculpted or merely happened by accident.

These things stay with you, and it remains a dominant ambition of mine to ask the Prime Minister, whoever he or she might be, possibly, I suppose, a gender-free transsexual any minute now, if that has not already happened, about how a rivet works. Of course, he or she will not have a clue and is, thus, in my definition, scarcely above illiteracy. I was thinking all of this during a visit to Spencer Lane-Jones in Warminster, a workshop specialising in Bristols.

Here, in my incongruous posh suit, I found myself engrossed and enchanted, talking to machinists and toolmakers with oily rags, who explained a castle nut (it's part of a secure fastening system), shims and machine-rolled threads. Just writing 'shims', I can again hear my father's voice. It's not a word used much in the Number 10 Policy Unit. And we are all so very much poorer for that.

Have any contact with Bristol and you start thinking about two things, besides the art of making. First, cars named after places. Lincoln does not count because the Ford children were thinking about something else, but Wartburg is a possible contender, even if it is a castle, not a town.

Personally, I like Wartburgs since in that demanding Cold War moment when Berliners ate cardboard and hugged chickens for warmth, the Ossis were required to use extraordinary ingenuity even to make a bucket, let alone a car. And since we have discussed the evocative power of olfactory sensation, half-burnt two-stroke mixture is comparable, for me, to a whiff of Chanel No. 5 on a cashmere scarf.

But the second thing you start thinking about when you start thinking Bristol is the links between the aircraft and car industries. In 1945, a lot of Swedish engineers found themselves sitting in the sauna with nothing to do but gloomily chew smørrebrød and dream of encounters with moose or reindeer, so we soon got Saab. It was the same in the West Country. The Bristol Aeroplane Company employed tens of thousands when demand for military equipment was high, but after the War, when demand was zero, or less, they decided to build a car.

Look at the rear of a Saab 92 or a Bristol 401 and there is a clear similarity in their tapered tails. Aerodynamics, lightweight materials and semi-monocoque structures were part of the aerospace inheritance. So too, were demandingly accurate panel fits.

Alas, I have not (yet) flown a 1916 Bristol F2 Fighter biplane, but I would not be surprised if a forties or fifties Bristol motor car accurately replicated many aspects of the dynamic experience. These cars, as reparations from war, acquired the rights to BMW 328 engines, but a turbine hum is not the sensation delivered. Instead, it is more a grinding symphony of loose-fitting parts, slightly out of tune. Progress is by surges, lurches and pauses, all accompanied by an overwhelming smell of petrol, now as rare as the evanescent scent of a Byzantine virgin's myrrh.

And why in 1951, when a Bristol 401 was perhaps the single most sophisticated car you could buy, Porsche included, was ergonomics still as little respected as dowsing or alien possession are today? You think about these things when you travel in an old Bristol. There is the smell and the thought of all those machine-rolled threads separating you from oblivion on the A36.

Your journey is a bargain between several types of forces, by no means all in sync, but it is not a journey imagined by a product planner or a marketeer. This journey was made possible by people who understood shims and rivets and made a car imbued with that understanding. When I think of the moribund perfection of my next car, I find this all impossibly, ineluctably, unforgettably romantic. Machines have life. Ask any Bristol owner.

CURVES

Car designers work like the army, or, in a perhaps less satisfactory comparison, the police. They are disciplined, but hermetic, existing in isolated communities and talking only to one another on a limited range of subjects in a specialised language intended to exclude the public. Yes, they travel, but only to places essentially identical wherever they are: grim military barracks or tragic international 'luxury' hotels. Of course, they all wear uniforms.

When they settle, it tends to be in quiet out-of-the-way places. This provinciality, I think, partly explains the deadly conservatism of the automobile industry as a whole. For example, I first met the very personable Thomas Ingenlath when he was working for Skoda in Mladá Boleslav, the Bohemian town whose Latin name is, most amusingly, Bumsla. I visited him again after he moved to Volvo, in Gothenburg.

I used to visit Sweden's second city often. My favourite memories include the restaurant where bear's paw was always on the menu, but always, sadly, 'off'. Second, very cautious nights in the Royal Batchelors' Club on Skyttegatan. Third, sitting on a rock in a swarm of summer midges drinking lingonberry and aquavit cocktails with Ingenlath's antique predecessor, Jan Wilsgaard, designer of the Amazon and the 240.

As a specific against suffocating provincial tedium, Thomas Ingenlath devised a design experiment that, so far as I am concerned, puts him, intellectually speaking, at the very front of his profession. Most designers still, and I am stifling a yawn here, have a Fender Stratocaster and a 1:24 model Ferrari or a carbon pushie in their studios for inspirational purposes. Instead, Ingenlath had sets of dry electrodes, which he strapped on to people's heads to measure spikes in neural activity when confronted with different aesthetic impulses. This is known not as torture, but as biofeedback.

Some of the results were not very surprising. Men prefer pictures of cars to pictures of crying babies. And fully two-thirds of women prefer good-looking men to beautiful cars. So far so unexceptional, but there is some interesting related data too.

Separate research shows that men like the front of cars, while women prefer the rears. I can hear Sigmund Freud's ghost hissing this is because the front suggests the male prerogative of penetration, while the rear suggests the anterior mating position favoured by most mammals (but not, generally, by members of the Royal Batchelors' Club).

But what Thomas Ingenlath was really after was an understanding of how curvaceous shapes affect us. Of course, Volvo's history includes the wonderful 240 Estate, possibly modelled on a forbidding Västergötland chieftain's coffin, but certainly the most rectilinear car ever made. Quite correctly, Ingenlath intuited that curves are more emotive than straight lines. And since he wanted future Volvo customers to emote heavily, the quest was, with the aid of dry electrodes and a scanner tuned into the brain's beta and gamma frequencies, to prove the allure of curves scientifically. Or why a 1951 Jaguar XK120 is more attractive than, say, Bill Towns' 1974 Lagonda.

Curves are also the subject of research by Oshin Vartanian at the University of Toronto. With dry electrodes all of his own, Professor Vartanian can show that our neural colonies all get excitedly up in arms when presented with a curve. Voluptuous shapes excite the part of the brain that processes emotion, while angular shapes stimulate the dark core of your amygdala, the brain's fear centre. But if you have seen a Lagonda, you know that.

I am not even sure you need dry electrodes to prove this. It is well understood that our reaction to curves is based on a simple associational response of an essentially sexual character. The world's most famous, if not best, woman architect was Zaha Hadid. She made her reputation with lascivious buttock-profile forms which are emphatically feminine. Her design for the Qatar 2022 FIFA World Cup reminded several critics of what we might decorously describe as the Delta of Venus. Meanwhile, the most famous classical sculpture of Venus is called *Venus Callipyge* (which means 'shapely bottom').

Love and curves go together and that's a relationship easily translated into the world of car design. For reasons based in the foundations of the brain's architecture, a curve, because it suggests either maternal warmth and well-being or an imminent erotic romp, touches a more profound part of the psyche than a parallelogram. Maybe this is because a woman's breasts are generally not right-angled.

Certainly, in architecture, Zaha Hadid's and Frank Gehry's curves currently get high approval, so Thomas Ingenlath was lockstepped with the Zeitgeist. And it just proves the aesthetician's belief that cars really are buildings on wheels.

THE FAMILY SALOON

'Family saloon' is not a term you will find often here, its twin suggestions of domestic duty and suburban routine being entirely at odds with the voluptuary tastes and aesthetic bent of most readers. Personally, I did not buy a car with four doors until my son was born. And after making symbolic tribute to the family in buying a saloon, I soon reverted to cramped and impractical two doors. My son and his sister were often driven to Rome while strapped to the luggage. Still, I have just seen a lovely Lancia Appia Series 3 for sale in California: here's a family saloon that would not make you feel you were claiming benefits.

But I am wondering if the idea of the 'family saloon' is disappearing. There is a lot of inertia in the motor industry, but for how long can automobile formats be retained after the social circumstances defining them cease to exist? Sure, politicians have gifted us with 'hard-working families' (and I wonder what a hard-working family saloon might look like), but most children are now born outside marriage and the conventional idea of a family is a memory more than a reality. 'Sports car'? Ideal for that top-down jaunt to the country pub! Does anybody really do that? 'Station wagon'? When did you ever pick anyone and their bags up from a train? We have taxis for that.

Right now, is a moment of extraordinary profligacy in the motor industry. It's what you might call an endgame paradox: as the historic moment of the heat-engine automobile nears its close, the variety of formats on offer to consumers multiplies insanely. There are two factors here, dangerously related: a lust for volume and an infatuation with the Chinese market, construed as an Asiatic El Dorado where infinite riches might easily be mined. Like El Dorado, that may soon prove a delusion.

The otherwise rational Germans are all demented by the pursuit of volume and have, so it sometimes seems, abandoned the principles of clarity and discipline which made their products attractive in the first place. Meanwhile, a slavish pandering to the Chinese market has perverted good sense. Instead of an engineering or design objective, because the Chinese demand it, BMW sells heavy, stretched 7 Series saloons entirely at odds with a distinguished dynamic tradition. Back in Europe, in a fine demonstration of corporate schizophrenia, BMW sells similar cars with either front or rear drive, perhaps for families who do not know whether they are coming or going. This from a manufacturer whose design traditions were once based in functionalist Bauhaus logic.

But the fickle Chinese will change their minds. Already, there are reports in the *New York Times* of once popular black Audis heartlessly abandoned and gathering toxic dust from Anqing to Zhangzhou. Rolls-Royce too seems to have underestimated the effect of internal Chinese austerity measures and sales have fallen dramatically. The Germans are, I think, going to be left with a lot of egg foo yung on their faces. And we will be left wondering what happened to all those cars we used to admire and enjoy.

The obscene profusion of the modern industry makes me feel nostalgic for the old formats and even for the family saloon, although I am very well aware that 'nostalgia' was originally defined as a psychosis. People suffering from nostalgia could only see pleasure in the past. That is certainly a shortcoming, but because the design brief for an old family saloon was absolutely clear – one engine, two front seats, three boxes, four doors – and its market uncomplicated, it inspired designers to create some of the loveliest cars ever.

Besides my Californian Lancia, I might add the Pininfarina Peugeot 404, the Michelotti-inspired Triumph 2000, the 'Neue Klasse' BMW 1500, the Fiat 1500, the Borgward Isabella, the Mercedes-Benz W123, the first generation aero Audi 100, the Citroën DS, naturally, and David Bache's glorious Rover P5 (which when cut down into a coupé brilliantly predicted the recent fashion for family saloons with a dramatic profile). These are some of the very best messengers of the automobile concept, being both beautiful and useful.

In the first paragraph I mentioned domestic duty and suburban routine as if they were demeaning, but they are the very context of the motor car. Le Corbusier, no weepy-eyed nostalgist, used to talk of cars being little houses on wheels. The family saloon projected an idea of domestic felicity on to the roads. And suburbia? This is where cars work best, since they are useless in big cities. And if you are in the wilderness, you need a military vehicle and an axe. Anyway, as with Frank Lloyd Wright (like most architects, a car man, who once owned a Mercedes-Benz gullwing), the family saloon helped make suburbia possible in the first place. This is where cars are at home.

As the endgame plays out, we will see many absurdities and enjoy several surprises. Perhaps the family saloon will be revived and the 'active tourer' or 'sports activity vehicle' consigned to the parts bin of history. I do rather hope so.

HYPERCARS

Obscenity has its own discipline, rules and conventions. You can do obscenity very well or very badly. So, it is impossible not to admire the latest category-bending, credibility-stretching, resource-destroying, patience-exhausting hypercars.

Never mind that 'hypercar' is a term coined twenty years ago by the thoughtful environmentalist Amory Lovins of the Rocky Mountain Institute, and that his intention was to describe a coming generation of ultra-light, intelligent, well-mannered economical cars. The hypercar name has been hijacked for thousand-horsepower demonstrations of atrocious redundancy. It's a phenomenon that future historians will see as redolent of our tense and uncomfortable moment in history.

I have my doubts about the established 'supercar', although this particular prefix at least suggests an aristocratic and confident superiority. But hyper means fanatical, rabid, excitable and excessive. Super means better. Hyper means worse.

Someone asked me if I thought the new hypercars were the T-Rex of the automobile, the last evolutionary spasm before extinction. I said I'd prefer to compare them to Hellenistic sculpture. When the Greeks had exhausted the Classical language of calm, serenity and good proportion, their plastic style became contorted, exaggerated and angry rather than beautiful.

Thus, the *Laocoön* of Polydoros, Agesander and Athenadoros in the Vatican Museum, a sculpture showing an old boy and his sons being most disagreeably strangled by malevolent sea serpents, was the exact equivalent in its day of the Lamborghini Veneno. It will always be remembered but never loved.

The PRs may say something different, but 'Veneno' sounds like venom. This is revealing not just of a nasty speciality of the Borgia court of the Renaissance, but of a horrible state of mind in the hypercar world. You might as well call a car Rapist, Contaminant, Sadist or Pathogen.

But watching a manufacturer working its way up a helix of bizarre excess is not uninteresting, even if it is repellent. True, a purist like me finds it ludicrous that Lamborghini with its zero credentials in racing culture should pretend to the ultimate in dynamics, but an avid pursuit of vulgarity is in its own way admirable.

Besides, there's another sort of performance involved with this $3.9-million, fractalised, atmosphere-bruising, hot and claustrophobic eco-catastrophe: its absurd price is not a limiting factor to its market, but an actual incentive. As Dylan sang, money doesn't talk, it swears.

At least the McLaren P1 reveals the attractive aspects of its inherited motorsport genes with astonishing negative acceleration of 2g under braking. And the active aero they won't allow in Formula One is all yours here for $1.3 million. There is a technical commitment in the McLaren specification which impresses, even if the result makes aesthetes and moralists wince.

But Ferrari is guilty of transgressions of taste too. 'LaFerrari' sounds like a grossly rouged someone with a push-up bra and no knickers you might meet in a Neapolitan bordello. I suppose this tells you all you need to know about the target market. *Due cento cinquanta gran turismo passo corto* sounds as lovely to my ear as the 250GT SWB is to my eye. If you ask me, Ferrari has lost something wonderful and found something horrible.

I wonder if technology, that great equaliser, will avenge itself on McLaren and Ferrari. Each has ancillary electric motors. Batteries for the Ferrari's motors are in a liquid-cooled box in front of the engine. This says a lot about heat management and its problems: LaFerrari must be as uncomfortably hot as our Neapolitan hooker sitting on a pizza oven at midday on Ferragosto.

Meanwhile, the McLaren's electric motor is powered by the same type of lithium-ion batteries which, being inclined to suffer thermal runaway leading to nuclear-strength blazes, have been so very troublesome in the Boeing 787. I am told you activate the electrics by pressing an Instant Power Assist button on the steering wheel. One day someone will do this and the dendrites in the battery will arc and the busy mini-Fukushima near your pale butt will turn your composite hypercar into a blaze of glory outside Harrods.

Would we snigger to see a smouldering P1 or a melted LaFerrari? That might be an inelegant response to divine retribution, but the public responds in kind to the sort of psychotic aggro that a hypercar transmits. When a look-at-me metallic purple Lamborghini Aventador was towed away in central London, the event was gleefully reported everywhere.

Popular hatred is not a good sign. No one ever wanted to punish the owner of a Lancia B20 or an XK120, let alone a dear MGA. But being a slave to excess is its own punishment. To which I'd only add: given the chance, I would like to see hypercar owners made to walk.

JOHANN WHO?

Something strange is happening in the industrial world. Consumer fatigue is mounting, demand is wilting, and rates of innovation are slowing: has there actually been a better office chair than Charles Eames' 1958 'Soft Pad' classic? I think not.

Meanwhile, entire product categories are disappearing. Still cameras, movie cameras, typewriters, hi-fi and landlines will soon be no more. Of course, it cannot be too long before the private car joins that melancholy list of redundancies. But curiously, while there is less work for designers to do, designers have never been more prominent.

And the most prominent of them all is Jony Ive, now ex-Apple. I often think that Jony is, in cash terms, the most valuable person in the world. At Apple he was in a gilded cage. Possibly surrounded by armed guards, and yet every designer aspires to his condition.

The word 'design' once described an activity, but in recent years has gone through a transformation so that it now means a commodity to be acquired, rather than a vocation to be pursued. In this way, designers have become trophies or talismans or fetishes. The presence of a mediagenic designer immediately confers value. Car designers now move about like footballers in the transfer window.

This churning of careers reveals a psychology where designers, aware of the malaise in their sector, see more security in personal advancement than in a lifetime of disciplined tenure. Unfortunately, it's lifetimes of disciplined tenure that tend to produce the best design.

This thread started me thinking about designer personalities and which might be the most valuable as a type. Harley Earl was the Jungian original: a huckster of brute genius who wore cinnamon or sky-blue coloured linen suits and lounged in Saarinen's incongruously Euro-normal General Motors Tech Center in Warren, Michigan, pointing out details on clay models with the well-polished toe of a crocodile loafer. He thought 'aluminium' had six syllables.

We have his equivalent today in Land Rover's Gerry McGovern. Dapper to a fault in waisted and sharply cut Savile Row schmutter, Gerry has big watches and big twinkly aspect. Never mind the cars, his presence alone has helped drag the brand out of the mud of utilitarianism into the coruscating universe of international catwalk luxury product. Of a similarly impressive specification is Mercedes-Benz's Gorden (sic) Wagener who has somehow persuaded a stern Supervisory Board in Swabia that he can operate best from surfer dude territory in California.

But I like arguing that design was at its most influential when designers were least well-known. At Mercedes-Benz, Bruno Sacco used to insist on 'continuity', not momentary sensation. And that applied to the cars as well as to the staff. Under Sacco was Johann Tomforde who interests me greatly because no one in this country seems ever to have heard of him, but his handwriting is all over the meticulous '82 C-Class and the superlative R129 SL of '89. He is also credited as the father of the epochal smart, one of the all-time great designs in any category. Of his SL, Tomforde said: 'This is a shape that won't bore you. In five years, it will still be desirable'. That's long-termism for you.

Let's take this investigation to Italy. What about Piero Castagnero at Lancia's Centro Stile who, as the company was already looking into the abyss, drew the utterly, utterly beautiful Fulvia Coupé? Castagnero was never well-known. But the supreme example of how great design evolves from continuity and commitment is the elegant Dante Giacosa. He created the original Topolino, its successor the Nuova Cinquecento, and his last car was the Fiat 128, technically interesting and aesthetically perfect. He describes in his autobiography *I miei quaranti anni di progetazzione alla Fiat* (1979) an enduring engagement with the company, its products and its culture. Forty years? That's what I call tenure.

And four decades gave Giacosa time to describe design in all its complexity: 'It comprehends the conception, working-out and final definition of the created object before it is materially executed. It is an imaginative process where development is made possible by the application of the sciences which constitute engineering, covering constructional science, mechanics, electrical engineering, electronics, and a wide field of knowledge including technology, ergonomics, commodity marketing etc. In the automobile field (as in architecture), aesthetics, biology and ergonomics are also extremely important'.

That's a very fair summary, but I would put it another way. Excellence in design comprises a balance between stability and evolution. You need a good idea to start with, but it must also be one capable of development over time. Job-hoppers can't do that. That's why I think the R129 Mercedes-Benz SL was supreme. At the time of its launch, so obviously new, but so obviously Mercedes. Pragmatic, refined, subtle, understated, impressive: the result of a settled and respected tradition.

Next time you see a car designer pirouetting before the media and doing rhetorical mash-ups, talking of BBQ surfacing, sun-dried swage lines, emotional intelligence and, Gawd-'elp-us 'DNA', remember what can be achieved by commitment, continuity and modesty.

And you'll be wondering about Johann Tomforde. He now runs a mobility consultancy in Böblingen. And he is thinking electricity.

RETROCAUSALITY

I imagine at the particle physics conferences where they meet, believers in retrocausality are asked to sit in a separate group so their foaming mouths, eye-rolling and terrible twitches do not distract the other delegates. While the maths makes it theoretically possible, retrocausality is not taken entirely seriously by orthodox scientists. It posits a belief that the future can influence the present and that the present can distort the past.

But this is not entirely mad. It's what car designers do all the time. Your sparkling new car is often two product cycles out of date: the designers have already designed the successor to its successor. They live in the future and work backwards to the present.

Those mid-term facelifts are part of a calculated plan to align an already imagined future closer to the present day. That fabulous old perfumed and pomaded white-shoed huckster Raymond – Studebaker Avanti – Loewy established the principle of MAYA. This stands for the Most Advanced, Yet Acceptable. Namely, let your imagination soar, then haul it back to the point where the consumer can tolerate it. Or buy its products.

But I sense things are changing. We are coming to a point where neither physics nor history seem linear or progressive. Or have you not seen the ads for the Fiat 124 Spider? They make explicit reference to its fifty-year-old predecessor whose shape it apes. The new car is, however, based on the Mazda MX-5, which itself was modelled on the 1963 Lotus Elan. Meanwhile, the new Land Rover Defender resembles its seventy-year-old grandfather.

Everything returns eventually: length of skirts, width of trousers, presence of facial hair, and television series. Confronted with a nouvelle cuisine horror of bamboo-smoked Argentinian oyster, with julienne of ortolan and a jus of Venus flytrap, or some such, the historian Jean-François Revel wailed 'who's to say if this is mad invention or a return to Roman cooking?' When, in 2014, Qantas took delivery of its first new Boeing 737-800, they had it painted in the old livery with Gert Sellheim's 1947 flying kanga on the tail. Of course, everyone loved it, even if the flight attendants were not in the period-correct, matching-numbers gorgeous Emilio Pucci frocks.

This is what's lazily called 'retro', but the very word deserves a little analysis because it is far from simple. A coinage of the troubled seventies, we have the French to blame. A *rétroviseur*, of course, means rear-view mirror, but our term probably comes from *rétrospectif*, a review of an artist's career. But retro also has a relationship to kitsch. And kitsch is best defined as the corpse that's left when anger leaves art.

Still, we can now see that a group of 'retro' cars that began with Nissan's 1987 Be-1 take on the Mini, continued with the 1989 S-Cargo, a Postmodern 2CV, developed through the BMW Z8 of 2000 and is still with us now in the BMW Mini, Fiat 500 and Ford Mustang, actually forms a coherent body of work with a language and vocabulary all its own.

Yet it is not mere copyism because who can say what exactly has been copied? The concept? Not really. The general arrangement? Hardly ever. The details? Actually, no. The character. C'mon! In most senses, these retro cars are, in fact, highly original. I mean, the sensibility that inspired the superb Citroën AZU camionnette was really not at all the same thing as the sensibility inspiring the silly S-Cargo.

Although there are Japanese precedents, it was J Mays' Audi Avus concept, which appeared at the 1991 Tokyo Motor Show, which is the key work in this movement. With clear references to prewar Auto Union racers, the Avus seeded the Audi TT and perhaps helped Mays persuade Volkswagen management to manufacture Concept One, which in 1997 turned into the reborn Beetle, whose historic Porsche source was possibly inspired by Josef Ganz's Maikäfer (May bug).

But how interesting it is that while the BMW Mini and Roberto Giolito's Fiat 500 have been successful, other retro cars have been less popular. With the revived Ford Thunderbird, J Mays' normally very sure hand lost its cunning. My wife saw a red one on King's Road. When I answered her question, she said 'No! Thunderbirds shouldn't be polite'. If you're going to mess with history, do as Hunter S Thompson said: 'buy the ticket, take the ride'. To avoid kitsch, forget politeness: you need to be angry. Or at least energetic.

Perhaps it's the idea of progress and novelty that's weirdly isolated in history. What a vain delusion it was that we might be able continuously to create newness. That was a Modern idea. Postmodernism revised that. And now we are post-Postmodern and in a terrible muddle. Again, I cite the Fiat 124 Spider.

In 1952, musician John Cage 'wrote' a famous piece called *4'33"*. The notation refers to the length of silence he recorded. You can listen to it any way you want. You can do that with history too. Cage later mused: 'The past must be invented/The future must be revised'. As I say, that's what designers do. It's just that some get more angry than others.

ANDY WARHOL

Was there anyone who better sensed the pulse of the late twentieth century than Andy Warhol? 'Iconic' is an abused term, but Warhol understood that popular culture had acquired an almost religious resonance. He did not have an artist's studio, he had The Factory with assembly lines – a place as significant to industrial New York as the Opera del Duomo to Renaissance Florence.

Warhol proposed closing down a department store so it could be preserved as a modern museum. He exploited celebrity, first of others, then, as it enlarged, his own. His career began as a commercial artist, drawing shoes. And then he became the most commercial of all artists. Making money, he once said, is the most beautiful thing of all. He wore a peroxide wig, aping his muse Marilyn. In a nightlife of flash photography, he was shot by a groupie-admirer.

And Warhol adored cars. In the Andy Warhol Museum retrospective in Pittsburgh, one of the earliest pictures was a drawing of his brother's delivery truck. A nervous disposition and a regime of substance abuse were deterrents to a driving licence, but Warhol's veneration of the automobile was sincere, even if it was often sardonic. With electric chairs, Hollywood, Elvis and supermarket packaging, they are recurrent in his iconography: his *Silver Car Crash*, of 1963, famously sold for $106.5 million.

A collaboration between austere Bavarian technocrats and the unchallenged leader of the Downtown Manhattan art, drugs and LGBT scene may seem incongruous, but it became one of history's happiest creative accidents, taking the collaborators places they might never before have imagined. BMW to the art-world vernissage, Warhol to Le Mans … at least in spirit.

Alas, there is no record of what was going on, or coming down, in The Factory when Warhol took the call, but transvestites shooting up and nudes wandering in thigh boots were not at all, one imagines, like a BMW Management Board meeting.

Warhol's 1979 M1 was only the fourth of BMW's Art Car programme, which began in 1975 with 3.0 CSLs painted by Alexander Calder, Frank Stella and Roy Lichtenstein. But not only was the car different, so was the artist's engagement: Calder, Stella and Lichtenstein had painted 1:5 scale models with their designs and these were implemented by a Munich paint shop. It was remote. But Warhol was personal.

Largely ignoring the brief that the Art Car could be anything at all provided it did not interfere with racing performance, Warhol first proposed flowers and camo, then a scheme which was all brown, including the windows. BMW demurred.

So, he decided to travel to Germany. None of the artists was paid for his work, but Warhol blagged the air tickets and installed himself and his entourage in the Vier Jahreszeiten Hotel, Munich's swishest. And then he set to work. Of course, Warhol had said that in future everyone would be famous for fifteen minutes. In the event, it took him just twenty-four minutes to create the most famous BMW ever. There he is with his decorator's paint and a three-inch brush, both heroic and absurd, intensely splashing away on a design of broad colour fields. If he was using sketches or a model, they are not obvious in the vintage video.

The intention, he intoned, was to suggest speed. Yet the beautiful paradox is this: in his gallery art Warhol wanted to reduce everything to the shiny status of mass production, uniform and characterless. But his BMW M1 is an autograph work by his own hand. You can see and feel the expressive texture of the brushwork, just as you can in a Rembrandt. And, as if to emphasise the specialness of this project, Warhol signed his name in the wet paint of the rear bumper, using his finger.

The car finished sixth at Le Mans and during the twenty-four hours some of the paint separated from the body, as if in a metaphor. But who wrote the rule that art must be functional? In 1951, the Museum of Modern Art had declared cars 'rolling sculpture'. Warhol had now created a racing picture. And BMW owns what is perhaps the most valuable car in the world.

CALIFORNIA

Roads do not, I think, get proper critical attention. I mean to say, it's not much use having a fine car if you do not have fine roads to drive it on. And none is finer than the amazing three-level interchange of Interstate-405 and Interstate-10 at Mar Vista, California, especially when seen from the air.

This, the junction of the Santa Monica and San Diego Freeways, is a marvel of sixties optimism and heroic concrete engineering. It was completed in 1964, the year the Beach Boys released *Fun, Fun, Fun* where fun was, I think quite correctly, identified with liberal and expressionistic use of daddy's Ford Thunderbird. By the lights of '64, this song could be the California State Anthem. Dear me, those surely were the days.

The Interchange was designed by Marilyn Jorgensen Reece who did more for women's causes than a whole rattling caboose of shrill feminists. Reece engineered gentle, sweeping curves so that T'birds with cart-spring suspension carrying Californian girls and boys could move at speed in an eroticised daze of sunshine and marijuana smoke. Indeed, there are challenging signs saying, 'You Must Enter at 55mph'.

On my first trip to California in 1980, 405/10 was an architectural pilgrimage site. I had been sent there by Reyner Banham, a founder of Pop Art and one of our greatest critics. Banham had written of the Interchange: 'A work of art, both as a pattern on the map, as a monument against the sky and as a kinetic experience'. I took in this spectacle then promptly drove up the Pacific Coast Highway in a Ford Mustang to pay my dues to Professor Banham at UCal Santa Cruz.

Banham died in 1988. Reece died in 2004 and the Interchange is now named after her, although I am not sure she would much care for it as a memorial. I was back in Los Angeles recently and revisited. Seen from the air is the only way you would want to contemplate the junction today because, experienced at road-level, it is a frightening, nihilistic nightmare. Forget that lilting Burt Bacharach song of Dionne Warwick's where she goes 'LA is a great big freeway'. No, it is not. LA is a great, big, strangulated near-death mess of stop-start traffic.

Entering the Freeway at 55mph, only viable if stoned, would be both murderous and suicidal since so far from being 'kinetic', it is an angry river of metal stretching to the horizon and, I am certain, beyond, possibly as far as Provo, Utah. As a driver, being here adds new richness to the semantics of torment. A blog I read while stationary declared 'I would rather drink a bowl of syphilis than be on the 405/10 Freeway at 5pm'. You really cannot say fairer than that.

I was in California with Charles March and a group of friends. The proprietor of Goodwood has another life as a photographer and there was a vernissage of his latest exhibition at a gallery downtown, once a senseless killing area, but now become a venue for art. Our gallerist was Adam Lindemann who notably sold a Jean-Michel Basquiat painting for $57.3 million. Lindemann also owns Andy Warhol's old estate at Montauk and is married to a daughter of Moshe Dayan, the dashing Israeli general with an eye-patch who was a hero of the 1967 Six-Day War. There's really never a dull moment if you are in Charles March's entourage.

So accordingly, we set off to rediscover the lost dream of motoring in California. Our venue was Jay Leno's Big Dog Garage in Burbank. After an hour from Beverly Hills in a lurching van, we found that dream realised indoors. Although Jay's enthusiasms are very well-known, nothing prepares you for the actual spectacle of vast industrial units with a madly eclectic range of cars, perfectly fettled and stretching almost to infinity, just like the traffic outside. Boggling? We need a bigger word. Among McLarens, Ferraris and Porsches of all types, I saw two Panhard PL17s, a '68 MG Midget, a delightful Honda S600, and when a metallic tangerine '63 Chrysler Turbine caught my eye, Jay obligingly started the engine for me. That is not an everyday experience, I can tell you.

But there was one car that captivated me. This was a coruscatingly immaculate '66 Oldsmobile Toronado, Bill Mitchell's last great shape and, I contest, the last great American car. But this being California, appearances deceived. Jay lifted the hood to reveal a Corvette C7R engine. Say 700bhp? And in this cetacean, power now went to the rear wheels. And these wheels were milled from solid drums of aluminium by Jay's own machinists, inspired by hot-rod culture and surely informed by the technical disciplines of the neighbouring Jet Propulsion Laboratory in Pasadena. You like an aerospace union and woven hoses? Here you have them.

I knew my choice was right because back in London I mentioned to J Mays that I had just returned from Big Dog and he said: 'Jeez! Did you see the Toronado?' It would be the ideal car for the 405/10 Interchange. But that's Californian dreaming, not a reality.

A JAPANESE DONKEY

My daughter has been to three universities. An old one, near J8 of the M40, and another in Paris. Then post-grad in London. She is now a restaurant anthropologist, a tribal leader in the capital's ferociously competitive catering jungle. She has a bar of her own and the *New York Times* recently called her our 'cocktail connoisseur'.

I mention this not out of annoying *amour propre*, although that is certainly a factor, but to establish her credentials as an acute observer of contemporary taste. We were together in Greece recently and looking out over the village from the terrace she said, 'Hey, pops, what's that really seriously cool car?'

It was a Suzuki Jimny, an automotive emoji. This version was the pick-up with tiny cab connected by stabilising bars to the flatbed. All the plastics were crazed and the glass opaque. It had a modest suspension lift and body panels in several different colours, all blasted matt by Aegean salt. The externally mounted spare was heroically rusted. I estimate that it had never been washed in twenty years. On this wild island, car washing is a wanton extravagance. They leave it to the *meltemi*, the evil wind that blows from Turkey.

Back in London, I am savouring the incongruity of writing about a crude Jimny on pages where sophisticated Italian sculpture is a more familiar sight, but let's try to understand the philosophical basis of the little Suzuki's unusually strong aesthetic appeal. To me, it is one of the great designs.

Most importantly, it's a question of size. In Sei Shōnagon's classic *Pillow Book* it says, 'all things small are beautiful'. That's not universally true and anyone who has seen a 1955 Suzuki Suzulight would agree, but like that little saloon, the Jimny's origins are in Japanese legislation. In 1949, a set of regulations was established to define a *keijidosha*: the most minimal car that was practicable. Engine capacity for what became known as 'kei cars' was originally restricted to 150cc, making Japan's urban speed limit of 40km/h a demented vision of inaccessible possibilities.

The first Jimny appeared in 1970. Legislation had by then been relaxed and it was powered by a 359cc two-stroke in-line twin. It is a pleasant imaginative exercise to consider the Godalmighty noise and reverberation this engine would have caused within the Jimny's construction of ribbed flat metal pressings. There would have been the flapping of canvas too. Jimny only acquired metal doors in 1979.

Japlish played its part in the car's birth as well. In 1970, topless bars in Tokyo were mistakenly called no-panty bars and Jimny's four-wheel-drive was called 'Allgrip'. The name itself is said to derive from a mishearing of the appellation 'Jimmy' by a Suzuki executive on a golf course in Scotland.

But the Japanese also have a cultural imperative towards miniaturisation. We all know about bonsai, that weird taming of nature. And the attractive ludic quality of the Jimny has its equivalent in *netsuke*, the tiny and very prized sculptures of animals which were originally used as toggles for clothing, thus, like the Suzuki, being both decorative and functional. Additionally, there is a precept in Zen that says smallness suggests power.

Then let's not forget military chic. While the cute Jimny could not easily be confused with the Sd.Kfz. 222 armoured car of Rommel's Afrika Korps, it is nonetheless dense with military signifiers. Its seventies contemporary was Sony's battlefield green CF-270L boom-box with emphatically chunky details and a DMZ-style whip aerial for radio reception when your disco cassettes jammed.

Moreover, in terms of design, the Jimny is a modest masterpiece in that a simple formula can be evolved into different formats: my daughter's cool pick-up, a neat tin-top and one with a big roll-bar and two canvas roofs, making an effect almost like a sedanca de ville. Unity-in-diversity is what the philosophers call this.

And while my friend, the philosophical Professor Simon May, has recently argued in a successful book that cuteness of the Hello Kitty sort is a worrying demonstration of the infantilism that's endemic to our troubled age, the endearing Suzuki Jimny accesses another, more profound level of our desire.

We are all through with arrogant and needless complexity. There are sub-menus on my Mercedes which I will not explore if I live to be one hundred and fifty. Meanwhile, the wine-list in my daughter's bar simply says Red, White, Pink or Fizz. That's your choice. Her generation is not at ease with fine wine, nor with fine cars. When she looks at a Jimny she senses something pure that is now lost. And, frankly, so do I.

And I almost forgot. While vehicle dynamics are scarcely a personal speciality, the Suzuki Jimny is a thigh-slapping hoot to drive. A little bit tippy with its high centre of gravity, underpowered and with mechanical connections in the steering and transmission being very haptic, you drive it with the intelligent anticipation needed in an old-fashioned sports car. On the road from Agnontas to Glossa, a Lamborghini would be embarrassed by a Jimny. That's another good reason to consider one.

CINEMA

Brigitte Bardot and Roger Vadim in St Tropez, 1956
Roger Vadim's 1956 film *Et Dieu ... créa la femme*, a
hymn to sun, sea and sex, built on the emerging cult of
St Tropez and established the reputation of Brigitte Bardot.
(She appears in a wet shirt.) Here the near collision in
St Tropez's Vieux Port uses the cars as proxies for sex.
Vadim is in a Lancia Aurelia B24, Bardot in a Simca. The
actress later lent her initials to the Ferrari BB512, one of
the most provocatively lascivious cars ever made.

Marilyn Monroe and Arthur Miller in their Ford T'bird, 1956
In Arthur Miller's 1949 play, *Death of a Salesman*, the sad
anti-hero Willy Loman treats the acquisition of a car as
no less than the complete realisation of The American
Dream. His psychic state is revealed by the car's condition.
In 1956, the new Mrs Arthur Miller, also known as Marilyn
Monroe, bought a raven-black Ford Thunderbird. It was
their regular ride around Roxbury, Connecticut. Just before
her death, she gave the car to the son of her acting coach,
Lee Strasberg.

Marcello Mastroianni on the set of *La Dolce Vita*, 1960
Cars play as important a role as the actors in Federico
Fellini's 1960 masterpiece, *La Dolce Vita*. A Ford
Thunderbird and a Ford Fairlane, louche *bargea*, are
suggestive of a refined cosmopolitanism. American cars
had something of a vogue in the Italy of that day: a
Chevrolet Corvette and a Cadillac also feature in the
film. But the other side of Mastroianni's 'Marcello Rubini'
character is revealed by the grumbly Triumph TR3 sports
car, driven by Mastroianni/Rubini throughout the film. The
actual car, 'un spider inglese', was imported to Italy in 1958
by a pasta tycoon.

MILTON WANKEL

Lost causes are fascinating, and this is why the NSU Ro80 and Milton Keynes make such compelling spectacles. 'There is no future for the future' essayist Max Beerbohm once defiantly explained to the Futurist pamphleteer, poet and activist FT Marinetti (who was so car-mad he suggested Venice should fill in its canals and turn them into motor roads).

The Ro80 and Milton Keynes are both fifty-year-old antiques, but they were conceived in a common delirium of futurist ambition. They have so much in common yet are so very different. One of the best descriptions of design is that it turns dumb materials into expressive forms. Here are expressive forms located at the polar extremes of aesthetics: a gorgeously sculpted car and an uncompromisingly rectilinear city.

So, how can it be that the two technologically infatuated cultures that simultaneously produced both the Ro80 and Milton Keynes produced such radically different artistic results?

Allow a vision to come into mind. One day in the mid-seventies, two young architects, Stuart Mosscrop and Christopher Woodward, set up a laser in the new Milton Keynes shopping building they had designed. As the enormous structure rose above the ground, Mosscrop and Woodward were (somewhat fanatically) keen to check that its astonishing one-kilometre-long horizontals were absolutely straight and true. No one has ever had a more fixed idea of what a building should be.

The two young architects had drunk deep at the source. Mosscrop, a blunt Yorkshireman, had worked in Chicago for skyscraper masters, Skidmore, Owings & Merrill, admiring while he was resident that city's fine patrimony of architecture by Mies van der Rohe, including the Lake Shore Drive Apartments and the Illinois Institute of Technology. Straight lines and fine details were here a sort of religion. Woodward had worked for Peter and Alison Smithson, the stern Modernists who had declared that architecture must be all about 'ordinariness and light'. Indeed.

Prime Minister Margaret Thatcher opened the Milton Keynes shopping centre on 25 September 1979. It is now Grade II-listed and, unusually for a shopping mall, since retail is a volatile and capricious culture, has been unchanged for more than forty years: a testament to the intelligence and effectiveness of the original design.

Yet Milton Keynes is routinely mocked. Francis Tibbalds, of the Royal Town Planning Institute, described the centre as 'bland, rigid, sterile and totally boring'. Others see it as beautiful, disciplined and idealistic. Even if there is a whiff of a failed Stalinist state as you leave the station to enter Central Milton Keynes, the effect on *Building Design*, the architects' trade paper, was – and remains – one of 'austere charm'.

And Milton Keynes was the last city designed for the automobile. It was designated a New Town on 23 January 1967. The idea was to avoid the centralised mistakes of the postwar New Towns, Stevenage, Peterlee and so on, and return to the idealistic origins of the Garden City Movement of Letchworth and Welwyn, but with many more roads and far better roundabouts. Milton Keynes is a city of separate, semi-autonomous communities and the roads run between them, not through them. In the year Milton Keynes was willed into existence by government fiat, NSU launched the Ro80 at the Frankfurt Motor Show. This was extraordinariness and delight.

How can two such disparate designs have shared the same historical origins? Each, by the standards of the day, was futuristic. Milton Keynes aimed to be the perfect modern city. The NSU Ro80 aimed to be the perfect modern car. If it were re-launched tomorrow, the car would still astonish us by its aesthetic and technical novelty. On the other hand, we will never build another Milton Keynes.

There is nothing about the NSU's specification or design that is confused, concessionary or second-hand. Its curves are gorgeous, its general arrangement bold, but refined. By intuition alone, the designer, Claus Luthe, created a body with a Cd of 0.355, the best of its day. The technical proposition was frankly boggling: its advanced Wankel twin-rotor engine made 113bhp from a nominal 995cc.

The glasshouse was huge and there was front-wheel-drive, unusual in a large car of its day. It was braked by Dunlop-ATE discs on all four wheels, in-board at the front to reduce unsprung weight. *Wunderbarlich*, there was a trick semi-automatic Fichtel & Sachs transmission: an electric switch operated the clutch as soon as pressure on the gear lever was detected. It was comfortable too. The seats were sumptuous and the interior extraordinarily spacious. The original *Autocar* road test said it was 'very advanced and a pleasure to drive'.

With proportions that are both interesting and flawless, the Ro80 still looks fresh and modern. The sole aesthetic error is wheels that do not quite fill the arches. Still, in its ten-year life, the only substantial change to the Ro80 was to move from rectangular to twin round headlamps in 1969.

To savour its radical proposition, remember that when the Ro80 was launched, you could still buy a Triumph Herald with its crude leaden lump of a push-rod motor, cartoon-origami architecture and murderous swing-axle rear suspension. By contrast, the Ro80 was a messenger from another and much more advanced civilisation.

This artistically organic and technically imaginative car had neither predecessors nor successors. It was completely original. In comparison, Milton Keynes had inspirational sources deep in architectural history. Yet while Milton Keynes flourishes, the Ro80 is almost forgotten. How can this be?

Trying to answer that question moves us into the philosophy of history. Art always reflects its own age, but sometimes that reflection can be confusing. If you look at old photographs of Mies van der Rohe's great buildings in fifties Chicago and New York, it is very curious how such formal architectural geometry is at odds with the cars parked outside it. The dissonance astonishes.

The Seagram Building in New York, so admired by the designers of the Milton Keynes shopping centre, is all exposed bronze I-beams, a building whose coordinates could be readily described on graph paper. And on the street outside, you see curvaceous Packards and Chevys with swooping fins in pursuit of another aesthetic entirely. Yet each was a contemporary product of the Zeitgeist. This troublesome term we get from Hegel, who said: 'no man can surpass his own time, for the spirit of his time is also his own spirit'. Maybe, but that spirit sometimes moves curiously.

Now, keeping in mind the image of the two young architects with their laser on a Buckinghamshire building site, consider another image from history. A young man, perhaps a little feckless, called Felix Wankel, joins the Nationalsozialistische Deutsche Arbeiterpartei in 1922. That clumping name (German Workers' National Socialist Party) was later abbreviated to 'Nazi'.

Although a keen hobbyist mechanic and tinkerer – an *Erfinder*, or inventor – the young Wankel was also an over-busy right-wing activist and enthusiastic anti-Semite. Between bouts of Jew-bashing and working as a printer's assistant, Wankel had dreamt of a new type of engine that would be 'half turbine, half reciprocating'. He had the general arrangement in mind as early as 1924.

Four years later, Wankel met Hitler and discussed his engine with the Führer. You can say one positive thing about Hitler, and that's he was alert to the symbolic and practical importance of engineering and design. Whatever Wankel was selling, he sold well. By 1931 he was a leader of the Hitler Youth in Baden, although he fell out with the local Gauleiter who was neither right-wing nor militaristic enough for Wankel's own tastes. Their row intensified and in 1933 Wankel was imprisoned, only to be sprung by Hitler, no less. In 1940, Felix Wankel, engine *Erfinder*, became an Obersturmbannführer in the SS.

Felix Wankel came, like many Nazis, from a humble background. He left school before his final examinations, the German Abitur, but apparently had an unusual spatial sense, often a function of untutored high intelligence. His engine was patented in 1929 and was designed to obviate the mechanical absurdities of the classic reciprocating engine where the pistons reach their maximum speed just micro-seconds before they are decelerated to standstill at top dead centre.

Instead, a triangular rotor worked smoothly and revved freely in an elliptical crankcase. There were fewer moving parts, fewer irrational decelerations. It was a beautiful idea, but there was no available technology to secure the integrity of the tip seals at each apex of the rotor. Today, titanium carbide, an ultra-strong metallo-ceramic composite, would create a seal with integrity, but no such thing was available to Wankel.

Consequently, Wankel engines always burnt oil and soon failed. This was ironic since Wankel himself spent the Second World War working on high-integrity seals for Kriegsmarine torpedoes and for Daimler-Benz and BMW. Still, by 1957, Wankel, under contract to NSU and working from a lab in his own home at Lindau on Lake Constance, had a credible working prototype of his engine. Eventually, Mazda, Daimler-Benz, Rolls-Royce, Alfa Romeo, Toyota, Ford and Suzuki all took out Wankel licences, but only the struggling NSU committed to production.

In 1963, the NSU Wankel-Spider, a cute, but awkward little thing, became the first-ever car with a rotary engine. NSU management now wanted a unique mid-size saloon with this same power source. Four years later, in August 1967, when Harold Wilson's government, in its white-heat-of-technology moment, gave the go-ahead to Milton Keynes, NSU launched the Ro80.

Claus Luthe was one of history's most extraordinary car designers, a man whose reputation is dimmed only by the atrocious fact that he murdered his drug-addled son. At Deutsche Fiat, he had been an influence on the design of the Nuova Cinquecento. He soon moved to NSU where the 1961 NSU Prinz 4 brought a little of Detroit to the *Kleinwagen* market. Luthe's inspiration here was Ned Nickels' Chevrolet Corvair, whose distinctive, sharp beltline also influenced the shapes of the BMW 1602, the Fiat 1300 and the Hillman Imp. The Ro80 too had a notable crease along its hips.

Delightful to think that this detail had its source in the Advanced Studio of the General Motors Tech Center in Warren, Michigan. These curious crosscurrents make the study of car design as absorbing as the mapping of global weather systems.

If there is any disappointment in the design of the Ro80, it is that the original and daring glass nose had to be abandoned because of legal requirements, although a memory of it appeared on the Citroën SM. Additionally, Luthe had originally proposed an instrument binnacle as advanced as the car's exterior. On cost grounds, this was replaced by an uninspired, incongruous and artless plank.

So, braced by the exhilarating linearity of Milton Keynes and muttering to myself *Nochmals die Senkrechte* (once again the vertical), as Bauhaus painter Paul Klee had once declared, I climbed aboard a Ro80, something I had been wanting to do since first seeing one as a boy.

Of course, fifty-year-old cars acquire a particular smell: that once sumptuous velour carries many olfactory ghosts. Still the chairs are very comfy and visibility perfect, although, ergonomically, you do feel you are sitting on the car rather than in it, but that's the swinging sixties for you.

There is a base-note of unburnt petrol and I began wondering if the very thin, black, shiny and rather large steering wheel actually had an aroma of its own. There is a choke, but the Wankel motor starts easily although by the standards of contemporary reciprocating engines, it does not seem especially smooth. Power delivery is, however, seamless and, if the three-speed gearbox graunches a bit, the lever's electric knob works much better than most devices that are half a century old. Ride is smooth and authoritative, although there is a lot of roll even at gentle cornering speeds. Still, it's a fine thing to know what the future used to feel like.

I was left wondering whether the Ro80 was the end of something or the beginning of something. TS Eliot came to my mind: 'Time present and time past/Are both perhaps present in time future'. The historical facts are confusing. In 1969, Volkswagen acquired NSU and merged it with Audi. Claus Luthe soon designed the very origami NSU K70 which then, as NSU died, became the Volkswagen K70, establishing that company's design language for the next decade. Then, Luthe's Audi 50 became the basis for the Volkswagen Polo. He left for BMW and drew the standard-setting 5 Series as well as the still dramatic 1989 8 Series.

Ten years after Mrs Thatcher opened the Milton Keynes shopping centre, she opened Terence Conran's and my Design Museum in Butler's Wharf. Stuart Mosscrop was also the architect. Some 37,398 NSU Ro80s were built in the car's ten-year life. Alas, there is not a single one of them in the museum's permanent collection, but if I had to choose one car that connects all strands of design – sculpture, technology, innovation, an expression of the Zeitgeist – into a single memorable entity, the Ro80 would be it.

As for the future, let's look back again to Eliot: 'In my beginning is my end. In succession/Houses rise and fall, crumble, are extended/Are removed, destroyed, restored'. Milton Keynes is doing fine. Ro80, RIP. 'What might have been is an abstraction/Remaining a perpetual possibility.' Poets always get it right.

STIRLING MOSS

Stirling Moss was a friend of a friend and I am fortunate to have had several small dinners with him. Everything you have heard is true. He was courteous, kindly and humorous. Of course, he had an accurate estimate of his own merits, but since in his chosen field he was one of the best there will ever be, only a sourpuss would deny him this indulgence. Curiously, he was unhappy flying. Not scared, of course, just anxious of ceding control of a machine to anyone else.

But what interests me most about the man is his presentation of self, his personal style. Stirling Moss had brand values long before the odious term became commonplace in the flipchart and lanyard community. It was something inherited. On his father's side they were Jews from the Rhineland, long assimilated here. But when the Ashkenazim immigrants – somewhat wild, tribal and exotic – arrived in the East End in the later nineteenth century, Stirling's grandfather Abraham anglicised the Biblical 'Moses' to 'Moss'.

And to emphasise the assimilation, Moss's father was called Alfred Ethelbert, as if he were a Saxon chieftain. A Scottish mother suggested 'Stirling' for Alf's son and a great brand was created: the name simultaneously suggests dynamism and stability with a top note of elegance. There is something about the cadence of 'Stirling Moss' that is absolutely correct. Nothing could detract from such extraordinary driving, but if he had been called Ken Brown we would all have been a little less star-struck.

Moss loved the camera and prepared for it. And cameras returned the compliment. From the beginning, he found elegant poses. Emotionally, he appears quite neutral in photographs: rarely exultant and rarely dejected. Evidently, he had personal control as finely tuned as his car control. In this last matter, Moss's style as a driver was completely distinctive: the head cocked slightly to one side, arms straight. Inimitable. He had an aesthetic.

And Moss accessorised himself with great care. Off the circuit, always well-cut suits. At the track, flatteringly form-fitting Aertex shirts. He had an identity bracelet when contemporary sportsmen were wearing leather boots and scratchy woollen underwear or using wooden racquets. I can imagine him now saying, in that strangely high rather lispy voice, 'Tachymetric Rim Scale'.

This was a feature of the best sports chronograph of his era: the Enicar Sherpa, so called because the manufacturer had sponsored the 1956 Swiss Everest expedition. Stirling appeared first in an Enicar advertisement in 1958.

Two years later he was giving his full endorsement: 'The Enicar Sherpa is definitely the watch I have always wanted'. They described it as a speedometer on your wrist. Later, Jim Clark wore one too.

But for me, Stirling's house was the most fascinating feature of his existence beyond motor-racing. While an exceptional human being, he was no great reader and, while highly intelligent, no intellectual (in the sense that an intellectual is someone who enjoys comparing and contrasting a wide range of ideas). But his house was one of the most advanced projects in the domestic architecture of its day. I have often wondered what his sources were because I do not imagine he spent much time scrutinising the *Architectural Review*.

The property – Stirling would have called it a 'pad' – was in Shepherd Street in London's Mayfair. Bought as a bombsite in 1961 for £5,000, he then spent £25,000 (about his annual earnings) on building a 2,500-square-foot, five-storey dwelling. Le Corbusier had said that a house should be a 'machine for living in' and this is what Stirling Moss made.

The bath could be operated by remote control. Over dinner, you pressed a button and the jacuzzi would be ready by the time you had finished your chicken Kiev and Black Forest gateau. The television and hi-fi were hidden behind secret doors. An electro-hydraulic dining table would rise and fall like an old cinema Wurlitzer. A hilariously wobbly device transported documents between him and his secretary. The facade had mirrored glass and, of course, there were compulsory Charles Eames chairs. Excited tabloids spoke of pink-dyed rabbit fur throws. Later, Williams F1 built Stirling a carbon-fibre lift. 'Quite a piece of gear', Stirling said. Alas, the car was absent one day and he fell down the lift shaft.

What was the inspiration? Was it Ken Adam's sets for James Bond's *Thunderball*, *Dr No* and *Goldfinger*? Can't have been: the chronology does not work. And while in the early sixties Archigram was experimenting with unrealised buildings which were environmental machinery, I really cannot see Stirling being influenced by radical architects with leftist views.

True, the great architectural historian Reyner Banham had an influential collaboration with the French illustrator François Dallegret, proposing similar ideas and enthusiastically advocating 'a baroque ensemble of domestic gadgets' surrounded by pipes and tubes. But that was 1965. Stirling Moss crossed the finishing-line before the radicals.

And then there was his personal transport. In Mayfair, a Vespa. Otherwise, he managed the astonishing graduation from Morris Minor to Jaguar XK120 to Facel Vega HK500. The latter he picked up in Paris and kept at Brussels Airport, using it exclusively for trips connecting the European Grands Prix.

I imagine Facel's wily Jean Daninos saw value in having Stirling Moss as an ambassador. He arranged for Picasso (who did not drive) to have an HK500 too. Stirling Moss and Pablo Picasso had much in common: unique geniuses who created identities as great as their gifts.

CULTURE

Carl Benz and family with the very first car, 1891
Carl Benz was the son of a train driver. His first ambition
was to be a locksmith, but soon diverted to iron foundries
and sheet-metal works. But 'gas engines' were his hobby
and soon they came to dominate his life … and later the
world. In 1882, he founded the Gasmotoren Fabrik
Mannheim and three years later filed his patent for the
ur-Auto. In 1888, his wife, Bertha, made the first ever car
journey: the 104 kilometres from Mannheim to Pforzheim.
She was visiting her mother.

Via Roma, Turin, 1964
The Alfa Romeo badge may incorporate Milan's coat of
arms, but there is no city more identified than Turin with
its most famous product: the Italian car. The 'T' in FIAT
stands for the city … Fabbrica Italiana Automobili Torino.
Historically, the presence of the royal stables, the
Cavalerizza Reale, designed by Benedetto Alfieri in 1740,
created an unusually heavy demand for fine carriages to
transport local royalty in suitable pomp and style. These
coachbuilders became the city's famous *carrozzerie*, now
designers as well as coachbuilders. When the world
turned from horse-drawn to horse-less carriages, they
were ready. Every car in this picture is Italian.

Rod Stewart with his latest Lamborghini, 1971
In 1971, Rod Stewart released his third and career-
defining album, *Every Picture Tells a Story*. To celebrate,
he bought his second Lamborghini Miura.

CARPORTS

Who coined the expression 'carport'? It was not the moustachioed and twinkly-eyed spec-builder of an executive estate in Reigate, but – according to architectural historian Ingrid Steffensen – Frank Lloyd Wright, the flamboyant genius of American architecture.

The relationship of cars with architecture is fascinating and profound. Roland Barthes's observation that 'cars today are our cathedrals' is just one aspect of it. Barthes meant to suggest that automobiles have the magical status and symbolic power of a great medieval church, at once familiar, but also mysterious.

Nikolaus Pevsner, who heroically catalogued *The Buildings of England* from the passenger seat while his wife drove, described cars as 'mobile controlled environments'. They are architecture on wheels. And like buildings, cars have plans, front and side elevations, with interiors carefully designed to enhance our moods. Besides, they betray our status and our yearnings.

To Wright, the car was the agent of two essential freedoms: mobility and personal expression. His landmark Robie House, on Chicago's South Woodlawn Avenue, had a three-car garage as early as 1908. Later, Wright insisted that his own Oak Park house should have dedicated petrol pumps. His personal transport in 1910 was the same model Stoddard-Dayton that won the first race at Indianapolis Motor Speedway the year before. Later, Wright progressed to a Cord L-29, a superlatively Modernist architectonic composition, and then to a pair of customised 1940 Lincoln Continentals.

Frank Lloyd Wright was not alone. Le Corbusier was no less than obsessed by cars. In his didactic book, *Vers une architecture* (1923), cars are presented as exemplars of good architectural design alongside Greek temples. His Voiture Minimum, of 1936, predicts the Citroën 2CV: a composition of absolute mechanical simplicity, based on Bauhaus principles of elemental geometry.

In Britain, by contrast, the settled opinion is that cars have had a damaging effect on both the man-made and natural environment. The Poet Laureate, John Betjeman, only had to say 'bypass' or 'dual carriageway' to excite sniggers of derision. When he said 'Cortina' it was coded language for the lower-middle class, which he so despised. But that the name was an eponym for an entire tribe attests to the car's semantic power. Be that as it may, a revisionist mood is upon us: in 2013, English Heritage listed thirteen car-related buildings.

Alas, they were too late to save my favourite: The Ox in Flames, on the Farnborough bypass in Kent, Britain's first drive-in. Opened in 1960 by a Minnesota entrepreneur called Marshall Reinig, The Ox in Flames offered experimental microwaved food to hungry travellers. The publicity photographs showed a well-favoured hostess with a very short skirt (four years before Mary Quant's mini) and, almost as attractive, a glorious two-tone Volkswagen Karmann Ghia. In the innocent years before The Beatles' first LP, this was precipitously advanced style.

Some of London's greatest twentieth-century buildings were created to serve the car. Chelsea's 1911 Michelin Building pioneered reinforced concrete and predicted the twenty-first century role of architecture as corporate identity. Piccadilly's favourite celebrity petting zoo, The Wolseley restaurant, was once a car showroom. Just across the road, Darracqs were sold on Bond Street. Wilkinson Eyre's Audi showroom, by the elevated section of the M4, is their twenty-first century equivalent. Outside the capital, Norman Foster's 1982 Renault Distribution Centre in Swindon confirmed his reputation as the high-priest of High-Tech. This he later confirmed with the astonishing 2004 McLaren Technology Centre, in Woking.

Or Ford's vast, redundant Dagenham plant seen from the air? There are few sights more expressive of the car's ambivalent place in practical life. When, in 1962, the enormous Park Lane garage was slipped under Hyde Park, it seemed noble and optimistic. Car parks are often elegiac. Alas the best of them all, Rodney Gordon's splendidly Brutalist 1969 Trinity Square Car Park in Gateshead, the other star in Michael Caine's *Get Carter*, was demolished in 2010.

But the very best monument to the car's romantic link to buildings is surely the Pennine Tower Restaurant on the M6, at Forton. Opened in 1965, before air travel had become an ordeal, its evocation of a control tower brought a certain exoticism to a motorway service area. I think of it as the architectural equivalent of its exact contemporary, the Ford Corsair: odd, useless, but wonderful. And perfectly redolent of its age.

Architecture is all about memories. That's something else it has in common with cars.

CAR SHOES

The great gourmet, Brillat-Savarin, had a famous saying: 'Tell me what you eat, and I will tell you what you are'. That can readily be adapted to: 'Tell me what you drive, and I will tell you what you want to be'. For me, cars have always been more than ends in themselves. I am just as much interested in the culture that surrounds them, the journeys, both real and metaphorical, that cars take you on.

This is why Michelin's *Guide Rouge* was so brilliant: it made the connection between cars and travel. And if you want people to travel (in order to use more tyres), what better inducement than to suggest wonderful hotels and restaurants to be enjoyed during their journey? That's a sort of magic that cannot be dispelled. You might be in a Dacia Logan and arriving at a Formula One hotel in a *zone industrielle* near Dunkirk, but in your imagination it is a Delahaye en route to the Hotel des Bains.

And tell me what you drive, and I will, additionally, tell you what you should be wearing. In its earliest days, driving was more an extreme sport than a reliable, rational form of travel, and proper protective clothing was required. When passengers and fuel lines were all exposed to the bitter cold, 'chauffeurs' (literally: heaters) were so called because it was their job to warm things up. Mr Toad gives reliable style pointers to this era. EH Shepard's illustrations to *The Wind in the Willows* show Mr Toad, a fierce early motoring enthusiast, in robust Harris tweed jacket and plus fours, a flat cap and a fur-lined coat.

Another pioneer of car-centric clothing was the splendid Edwardian character Dorothy Levitt who, aged twenty-three, manhandled her 8bhp de Dion Bouton from London to Liverpool and back in a brisk two days (something you'd be hard-pressed to repeat today). Later, brisker still, she became known as 'the Fastest Girl on Earth'. She always travelled with a Pomeranian dog and a recoil-less Colt Automatic as companions and, in her 1909 book *The Woman and the Car*, published the year after *Willows*, she has valuable advice about clothing. Beware, she says, of leather because it does not wear well. Also, the modern motoring woman should avoid 'lace or fluffy adjuncts to [her] toilet'. Instead, like Mr Toad, a sturdy, fur-lined tweed coat is recommended.

As cars became more civilised, elective elegance – as opposed to survival-mandated protection – began to enter the range of possibilities in the driver's wardrobe. In 1927, Isadora Duncan was undone when her elaborate scarf became trapped in the wheels of her Amilcar while travelling along the

Promenade des Anglais. 'Affectations can be dangerous', commented Gertrude Stein acidly. Two years later, Tamara de Lempicka painted her famous self-portrait, an image by which the age of Art Deco will forever be remembered. She, the Femme Deco, is in a green Bugatti with a natty hat by Hermès.

So that was the car hat. Then there evolved the car coat and, a little later, the car shoe. I think Mr Toad and Ms Levitt (she never married, and was found dead of a morphine overdose, aged forty) can claim responsibility for the invention of the former, although its design was also influenced by the Jeep coats used by servicemen in the Second World War. A car coat is cut to mid-thigh length since that leaves the feet unencumbered to operate pedals. And it became a staple of the male wardrobe in the sixties, the sheepskin variant being especially popular in a certain milieu that also drank gin and tonic. In this, the Age of Aquarius, there were even more opulent options than shearling; one furrier, missing the opportunity for tasteless jokes about pussy, advertising the possibility of enjoying 'ocelot in your Jaguar'.

No one, I think, wears car coats nowadays, not even ironically, but the car shoe has evolved into an enduring design classic. It was in 1963 that Gianni Mostile had the happy idea of combining the moccasin of the Algonquin Powhatan Native Americans, a slipper-like garment of soft leather, with rubber studs on the sole inspired, perhaps, by Pirelli and a perceived need for high coefficients of friction. Certainly, the original logo of Mostile's Car Shoe company included a tyre. The car shoe has now become an often-imitated generic, and *gommini* (literally: rubbers) are in the uniform of a twenty-first century fashion crowd who do not, generally, drive.

There's a nice absurdity therein, but car shoes are most pleasing as messengers from an earlier moment in history when an immaculately groomed Italian gentleman would think it barbaric to wear heavy and inflexible 'walking' shoes to control the feather-delicate pedals of his Alfa or his Lancia. You need a proper driving shoe. It is upon such exquisite refinements that civilisation was built. The car coat and hat might have gone the way of the hand crank, but car shoes remain as a small, but perfect, reminder of the car's contribution to culture as a whole.

IT WAS A DATE

The superlatively imperious *Cahiers du Cinéma*, a Paris journal that treats movies (especially French movies) with the ultramontane reverence you'd expect if Moses were the cinematographer, was very sniffy about the pied-noir auteur, Claude Lelouch. Reviewing one of his films, the *Cahiers* said, 'Remember the name Claude Lelouch … because you are not going to hear it again'. Bof! Another chapter is written in the glorious history of French *snobbisme*. Except, as we will see, this was not really true.

Lelouch had a nice affinity for the automobile. His 1966 *Un Homme et Une Femme* starred a Ford Mustang motoring purposefully through Normandy mist on a romantic assignment, the V8's own grumbly music in nice contrast to Francis Lai's tinkly jazz-pop soundtrack (as irritatingly memorable as the ur-Nokia ringtone).

But the picture I like best is a black-and-white still, taken in August 1976, which shows Lelouch in contortions as he attaches a 35mm movie camera to a rigid frame mounted just ahead of the radiator of his own Mercedes-Benz 450SEL 6.9. This was no ordinary Mercedes. The 6.9's engine had fighter-aircraft-style sodium-filled valves and was the first car to offer the now universal Anti-lock Brake System, or what we call ABS. Its optional hydropneumatic suspension allowed this ship-sized vehicle to be cornered like a Mini, but Lelouch found another benefit in the high-pressure system: it created a very steady camera platform.

He had just finished filming *Si c'était à refaire* with Catherine Deneuve. That, significant for this story, translates as 'If I had to do it again'. With 300 metres of valuable stock left over, Lelouch had decided to indulge a personal whim. This led to one of the masterpieces of the cinéma-vérité school which, in turn, has become the ultimate car film. The short is called *C'était un rendez-vous* and it was first screened late in 1976 at an event which is said to have led to the author's arrest on account of fabulous and wholly admissible evidence that, in the making of it, traffic laws, public morals and common sense had been outrageously debauched for nine sensational minutes. Yet, paradoxically, Lelouch's famous car film does not show the car that made it possible.

His whim was a simple one: film a drive across Paris at dawn at maximum practicable speed, irrespective of the law. Lelouch left the Porte Dauphine at 05:30 and arrived at the Basilica of Sacré-Coeur just before 05:40. The film has no edits, and the effect is mesmerising. In its way, *C'était* is similar to JMW

Turner's great 1842 painting, now in Tate Britain, *Snow Storm: Steam-Boat off a Harbour's Mouth*, which resulted from the artist being strapped to the mast for four hours so he could experience the full, unblinking horror of it all and then retell it.

Because the car is never seen, there were soon rumours that it was a Ferrari 275 GTB and the driver was the Grand Prix *pilote* Jacques Laffite. But the driver was Lelouch himself and the Ferrari soundtrack was dubbed after the event since, no matter what its other virtues, the three-speed auto Mercedes was low on acoustic drama. Still, certain risks were taken as the stately limo sped through the City of Light as the sun came up.

Lelouch told the documentary maker Richard Symons: 'When you get to a red light at 150-200km/h, if you don't see anything to the right or the left, then nothing is coming. For there to be a risk, there would have to be another crazy person doing this at the same speed'. (This reminds me of the time the distinguished graphic designer Marcello Minale told me his own interesting theory about motorway driving. Since, Minale argued, no police car was as fast as his Ferrari, if he set out at the bottom of the M1 and drove flat-out, no one could catch him.)

From the Porte Dauphine, perhaps with the intention of testing the Minale Theory, Lelouch drove up Avenue Foch, past l'Arc de Triomphe, down the Champs-Élysées and on to the Place de la Concorde. Without hesitation, still less stopping, he continued past the Tuileries to the Place du Carrousel, up the Avenue de l'Opéra and then, with the camera still running, wiggled through the backstreets until the red-light area of Pigalle. Then up the Boulevard de Clichy, more backstreets (with one duff turn requiring some reversing) and high up into Montmartre. The film, with its swaggering ego and scant regard for what we now call Health & Safety, was 'very symbolic of my life', Lelouch said. 'It's a totally immoral film, but I am glad it exists.'

Impossible not to agree. Only the dullest person could fail to respond without a back-of-the-neck prickle to the wailing engine note, the blurred Paris landmarks, and feel a sense of 'I wish I'd had the nerve …' And do you know what's most mysterious about *C'était un rendez-vous*? For all the outrage and piracy it entailed, all the cultish reverence it generated, the journey was, in fact, rather sedate. The route was a fraction less than ten kilometres and the average speed was 77km/h or, say, a little less than fifty miles per hour. On empty roads. So, really no big deal.

Remember this: the great beauty and fascination of cars is independent of speed.

EXHAUST

A threnody is a hymn of mourning, of lament. Not necessarily depressing like a dirge, threnodies can be stirring and stimulating. The famous aria in Purcell's *Dido and Aeneas* (with its unforgettable jaunty refrain 'when I am laid in earth') is a good example.

And the howling of fatally injured animals can be thrilling too, as Edmund Burke pointed out in his 1757 treatise on the Sublime and the Beautiful. This is what I was thinking while watching a recording of a recent Grand Prix. It was on a laptop so the sound quality was poor, but I am privileged long ago to have heard a Ferrari Formula One engine flat-out and that was what I was hearing in my memory as I saw a driver lose power in the race and creep to retirement. Sure, it was a rallentando as the engine died, a noise fading to silence but beautiful nonetheless. A fatally injured animal, indeed.

When the problem was diagnosed as a melted piston, I felt a strong pang of nostalgia. Melted pistons, dropped valves, blown gaskets, thrown con-rods and shattered crankshafts were the tribute Nature demanded for such glorious aural performance. I wish Professor Otto could have witnessed the final stages of his experiment.

As the Age of Combustion approaches the end of its long decline, collapsing with it will be status, pride, excitement, speed, fear and a lot of really bad stuff as well. But there is something delicious to be enjoyed at the end of any era. I bet Pompeii in AD 78 was most amusing with its flourishing hot food shops, taverns and bordellos (Latin spoken).

Right now, we are, cosmically speaking, in the Stelliferous Era, where all matter is arranged in the stars and galaxies. But soon, the astrophysicists say, all the bright stars will be gone, their energy exhausted. I feel the same way at the demise of the combustion engine.

I'll miss the noise. A rattly diesel taxi outside at dawn, always the overture to an interesting trip abroad. My father's MGB; people forget now how characterful that crude B-series four sounded. A loud American V8 (the 'mating song of the asshole' someone once said). The whining fan of an air-cooled Porsche or the amiable whuffle of a Deux Chevaux. Professor Otto's controlled explosions certainly created personality.

Remember that before the Industrial Revolution, the world was quiet. And boring. The publisher of America's first newspaper said he would print every month, whether there was anything to report or not.

There is a wonderful untranslatable German expression, *Sehnsucht*, referring to that lovely, but disturbing, sense that something is missing in our lives. Isn't *Sehnsucht* what we feel when we look at classic car ads? How might our lives be improved by a burgundy 1952 DB2 with painted wire wheels and a popping exhaust?

In 2008, the Hiscox insurance company commissioned a study about engine noise. Every woman tested in the study was found to have an increased level of testosterone, which is to say an enhanced sexual appetite, after hearing a Maserati. Equally, it was found that the sound of an ordinary engine had the opposite effect, diminishing the sex hormone. Imagine, if you dare, what the silence of an electric motor might do to the libido.

It is time to ask: when did you last see someone moved to tears by a battery? (Unless it was a Tesla customer, watching his car combust as the lithium-ion cells demonstrated the dramatic effects of thermal runaway.)

For all its liberating influence, let's accept that the car also enslaved. John Dos Passos described Ford's production-line in his 1936 fable, *The Big Money*: 'Production was improving all the time; less waste, more spotters, strawbosses, stool-pigeons (fifteen minutes for lunch, three minutes to go to the toilet … reach under, adjust washer, screw down bolt, shove in cotter pin, reachunder, adjustwasher, screwdown bolt … until every ounce of life was sucked off into production and at night the workmen went home grey shaking husks).'

Now everyone senses a moment of crisis: environmental anxiety, new energy sources, trade wars, changing travel habits, 6G, Artificial Intelligence and consumer fatigue (especially in cities) will all alter our assumptions about and expectations of the private car and, indeed, of travel itself. In some areas, ownership of a car is already stigmatised. To say, 'I have got a Porsche' is, in sensitive enclaves, similar to saying 'I have a criminal record'.

The manual and intellectual efforts required to drive a Porsche effectively will be replaced by electronic driver aids. You probably know the sort of thing: lane-change warnings, parking sensors, self-parking. These have existed as separate systems for many years now. They just need to get stitched together by 6G networks and sat-nav needs to get accurate to inches rather than quarter miles and – SHAZAM – we will have the promised autonomous cars.

Does anyone today remember driving gloves? They solemnised the rite of driving, adding tactile delicacy to the matter of holding the steering wheel, the prime interface between man and machine. Mr Toad had, as I recall, a deerstalker and goggles too. People used to dress up to 'go for a drive' just as people used to dress up to get on a plane. But to be doleful about the future of travel itself: the jet plane promised to democratise the luxury of air travel but instead it universalised mediocrity, spreading suburbanism across the planet.

Meanwhile, combustion-engine cars are entering their endgame paradox: as they become less useful in congested cities and thwarted by eco-angst and legislation on the open road, their specification and performance are uselessly enhanced.

Perhaps it is true that the mark of an advanced society is not that the poor have cars, but that the rich use public transport. Certainly, mobility will no longer be limited to or defined by the idea of a road trip. Once you felt free if you owned a car, now you feel free if you don't.

Macrocosmically, globalisation and over-tourism have destroyed destinations making one version of travel futile. A road trip today is a harrowing ordeal, not a romantic adventure. Microcosmically, the use of a private car in a city ceased to be sensible about twenty years ago. In any case, geo-fenced grocery deliveries by robo-cars will mean you never have to drive to Sainsbury's again. Unless, that is, you might actually want to.

Why? Because the great truth of consumer behaviour is that rationality plays no part in it. Cars became popular for reasons sunk very deep in the unventilated pools of the human psyche. They say the autonomous electric car will set us free. But the driverless car will, in every sense, lack soul. And people like soul, even if it is their prison warder.

And autonomy ignores the psychological reality of car ownership which is, if we are honest, based on concepts of pride and prowess and personality. Having Google drive you home when you have been over-served at a party might be a Health & Safety convenience, but who would want Google to determine the vectors of a romantic road trip? A driverless car will be no more involving than using an ATM.

The debate about the future of the car is as much social and cultural as technical. Of course (partial) electrification and (partial) autonomy are already realities and these will in future affect the design of cities and human behaviour as much as they affect the design of the car itself.

Our horseless carriage carries assumptions about mobility based in the nineteenth century. Ponder, for example, all the various meanings of 'steering wheel', that thing you caressed with your driving gloves: it speaks of authority, of an imperial mentality, of enthronement, of a personal destiny directed by the deft manual movements of a sole proprietor. Soon, the steering wheel will disappear. But what will take its place?

'The screen will replace the steering wheel. You'll watch the news as you travel. Or movies.' So says Norman Foster, an architect so committed to a tech future that he has designed a glass 'infinite loop' for Apple at its Cupertino HQ in California.

In theory you could eat and sleep in your autonomous car. If you did, your home might not need so many bedrooms nor so large a kitchen. But this assumes you are still a commuter. Of this I am not convinced.

The evolution of the steering wheel tracks the arc of the car itself, from Bertha Benz's ungainly nineteenth-century contraption to today's decadent, flawless, obsolete indulgences. Before the wheel there was a tiller, a hangover from the horse-drawn carriage. The 1898 Panhard was the very first car where a circular wheel replaced a horizontal bar. In the *Flash Gordon* era, American manufacturers experimented unsuccessfully with left-right wrist hoops for guidance, a little like an Airbus's fly-by-wire sidesticks.

Lately, the steering wheel has assumed greater technical significance than mere guidance of the front wheels: other functions have, following the early arrival of a push-button for the electric horn, migrated there. It has become a demonstration of ergonomics with soothing textures, satisfying radii and meaningful bulges, all to assist the satisfying mechanical craft of changing direction. At least as determined by a humanoid.

In a Formula One racing car, the circular wheel has evolved into a foam-covered trapezoid that's a mission control centre. A racing driver monitors a wheel-bound LCD screen for real-time data, using tiny knobs and buttons to adjust – with his gloved hands – the differential settings, air-fuel mix, brake balance and miscellaneous configurable performance maps.

A generation ago we might excitedly have predicted that this portfolio of techno-porn would transfer to road cars, but this will not happen because, when it arrives, the fully autonomous car will be guided by other means. The steering wheel will go the way of the tiller. Satellite signals from outer space, car-based radar and camera data will replace manual inputs in the matter of steering. The steering wheel will become as redundant as driving gloves. And designers are pondering what to do next.

When compact electric motors replace huge hot metal lumps, the fundamentals of a car's appearance will be changed; the autonomous vehicle need no longer be dedicated to an individual driver and his or her filthy, oil-based erotic fantasies. The autonomous vehicle can be anything. And when anything goes, very often nothing does.

If this happens, it will change the way we live. Just as cars once created suburbs, they will now change the shape of cities. Artificial Intelligence will mean there is no future need for traffic signals, directional signs or any of the ugly clutter of street furniture. Sat-nav has already caused the demise of paper maps. The print edition of Michelin's *Guide Rouge* can surely, alas, not survive much longer. Crash barriers will not be needed because your intelligent car will have made a moral decision not to mount the pavement and obliterate pedestrians. A human driver makes a million minor unconscious moral judgements in any journey. Run an amber light? Park on the kerb? No problem. A machine can make more judgements more quickly than you can.

But there are cultural problems here beyond the scope of technology. In 2018, *Nature* published the largest-to-date study of machine ethics, in conjunction with MIT's Media Lab. The survey showed that, in an impending collision, it was a general human response to spare humans over pets and to favour individuals over groups. And in East Asian cultures with a legacy of Confucianism, there was, if an accident was approaching fast, a tendency to favour preserving the old rather the young because that's the way Confucians reckon things. You would not want your intelligent car with code written in Shanghai's Jiao Tong University to drive itself anywhere near your kids' school.

What will robo-cars do for us? Robo-cars used for ridesharing will be in continuous use. Therefore, the need for city-centre parking will disappear. By some estimates, about thirty per cent of central Manchester, for example, is presently devoted to surface parking: that valuable land can now be put to better use. Meanwhile, people will no longer want to go on trips to the country to see cows and fields because urban farms (and green residential towers inspired by Stefano Boeri's Bosco Verticale in Milan) will blur the distinction between town and country.

Then there is the matter of materials. What's new? The much satirised Soviet-era Trabant had a body made of Duraplast, a hard plastic made from recycled cotton waste from the USSR, and phenol resin salvaged from the filthy DDR dye industry. Admittedly, it looked like a cheap suitcase, but the BMW i3, by consent the most complete and intelligently thought-out full-electric car yet to go to market, is made of CFRP (carbon fibre reinforced plastic). Its interior uses a mash-up of bamboo.

But, even as (almost) everyone predicts the demise of Professor Otto's engine, power sources remain the nagging question for the car of the future. When Credit Suisse predicts that by 2040 there will be an annual demand from the world's population of electric cars for 3.7 TWh (that's terawatt-hours) of battery life requiring a hundred Tesla-type 'gigafactories', the ones they have in Reno, consuming three million tons of lithium; well, when Credit Suisse predicts this, we can be very glad that there may possibly be life left in the old dog that is Internal Combustion.

Claims for electrical virtue are exaggerated where they are not completely misunderstood. Some predict it will be twenty years before the charge-density of batteries makes electrical autonomy viable. In any case, all those millions of tons of lithium come from territories economically colonised by China, thus, supply is under the whim of a Communist dictator.

Additionally, a 'clean' electric car uses electricity acquired from coal, gas or nuclear sources. And while hydrogen-powered fuel cells work elegantly, exhausting only water as waste, hydrogen is expensive to transport and store, at the same time requiring enormous amounts of energy to compress the gas into practical volumes.

John Heywood is Professor of Mechanical Engineering at MIT. His is a rogue proposition, but he nonetheless predicts that by 2050, sixty per cent of cars and light commercials will *still* use internal combustion. He says only fifteen per cent will be fully electric … the real gains in vehicle economy will come from downsizing and better aerodynamics. Besides, a visit to a petrol filling-station allows you to transfer 10MW of energy in less than five minutes. Achieving that efficiency with a Tesla would require a cable too big to hold. When Professor Heywood asked himself if, in future, he should be teaching his students about Professor Otto's internal combustion or sustainable electrochemistry, he answered: 'Both!'

Whatever happens, it seems the semantics of the car will no longer be tightly focused on selfish notions of enthronement, ownership and dominance, but a utopian one of shared space. Cars may become ever more like mobile architecture: spaces to be enjoyed. But what will replace the fascination and romance? Well-designed screens will have their allure, but the thirst for style, power and control may not be easily quenched.

Norman Foster concludes: 'And when you arrive, after a pacific journey staring at a screen, you'll get into your classic car and indulge yourself on the track'.

JG Ballard's prediction was, perhaps, not so very wrong.

INDEX

Page numbers in *italics* refer to illustrations

ACKNOWLEDGMENTS

This is a selection, of my own favourites, of articles I have written for *Octane*, the best classic car magazine in the world. So, thanks first to the founding editor, Robert – Shaguar – Coucher, who more than a decade ago had the whimsical notion of giving me a column called 'The Aesthete'. I don't think either of us believed it would last so long. There's a wonderfully good-natured collegiality about *Octane*, perhaps a sort of Blitz Spirit created by man's most ingenious invention being under attack on so many fronts. Huge thanks to the current editorial First XI who made this possible: James Elliott, Mark Dixon and Glen Waddington.

Two long-term motor industry insiders have over the years been extremely generous with introductions, privileged insights and what we call 'facilities', each helping me to understand the glorious and absurd complexities of their business. The first is John Southgate, Ford's devilishly subtle and accomplished one-time PR boss. Cheerfully retired to a South Australian sheep farm, which he patrols in a V8 ute, John knows at least as much about fine wine as he does about fine cars. The second is Richard Gadeselli, for a long time FIAT's equivalent. He performs the impressive double act of appearing Italian to the English, and English to the Italians, while talking with equal passion in each language about a Cinquecento or a Superfast.

Then there is the group of friends whose knowledge and opinions are of such value that I cannot resist stealing them. Gavin Green, last of the long line of Australians who did a Heimlich manoeuvre on British journalism. Stefano Pasini, an eye surgeon in Bologna, Lamborghini expert, and possessor of encyclopaedic (and ironic) knowledge of all things Italian. Christopher Butt, in Hamburg, is the best critic of car design I know. In New York, and elsewhere, Carl Magnusson has a droll take on everything that interests me. Locally, Andrew Nahum and Jonathan Glancey always generously answer my questions. And then there is Patrick Uden, the most meticulous machine romantic I know, a collector of ALPA cameras and a restorer of Citroëns. Once you have changed the wheel of a knicker-pink Ford Thunderbird on Manhattan's 44th Street in rush hour, as we did in 1980, you remain friends for life.

Perhaps because their work is so articulate, often car designers have little sensible to put in words. But there are exceptions. J Mays has intelligence and charm in spades and is free in spreading them. Patrick Le Quément has the sagesse of a Sorbonne intello, while Ian Cameron has a possibly unique career portfolio, including Pininfarina, IVECO trucks and Rolls-Royce. Then there is Flavio Manzoni of Ferrari, elegant in thought, dress and deed. Some of my best insights come from these four.

Not at all least, but (almost) last is the publisher, David Jenkins. We are both Welshmen and share an indifference to conventional pieties. He also insists, against all fashion, on producing beautiful books, assisted here by his excellent designer, Jean-Michel Dentand.

Last, my late father, Donald Sydney Staines Bayley. He would have been proud of *The Age of Combustion*, although he was more inclined to read menus than books. Nevertheless, he inspired my continuing belief that owning a fine car is a necessary part of civilised life. But maybe not for much longer!

CREDITS

Cover: imageBROKER/Alamy Stock Photo; pp10-1: Popperfoto/Getty Images; pp12-3: Steve Wood/Daily Express/Getty Images; pp14-5: Paul Popper/Popperfoto/ Getty Images; pp26-7: Sjoerd van der Wal/Getty Images; pp28-9: Bettmann/Getty Images, pp30-1: Bob D'Olivo/ The Enthusiast Network/Getty Images; pp50-1: Library of Congress/Corbis/VCG/Getty Images; pp52-3: Keystone-France/Gamma-Keystone/Getty Images; pp54-5: Walter Sanders/The LIFE Picture Collection/ Getty Images; pp90-1: Topical Press Agency/Hulton Archive/Getty Images; pp92-3: Los Angeles Examiner/ USC Libraries/Corbis/Getty Images; pp94-5: Staff/ Mirrorpix/Getty Images; pp114-5: Bettmann/Getty Images; pp116-7: John Dominis/The LIFE Picture Collection/Getty Images; pp118-9: Bettmann/Getty Images; pp154-5: National Motor Museum/Heritage Images/Getty Images; pp156-7: Bettmann/Getty Images; pp158-9: Touring Club Italiano/Marka/Universal Images Group/Getty Images; pp186-7: Michou Simon/Paris Match Archive/Getty Images; pp188-9: Paul Schutzer/The LIFE Images Collection/Getty Images; pp190-1: Pierluigi Praturlon/ Reporters Associati & Archivi/Mondadori Portfolio/Getty Images; pp202-3: Bettmann/Getty Images; pp204-5: Sergio Del Grande; Giorgio Lotti/Mondadori/Getty Images; pp206-7: Victor Blackman/Express/Hulton Archive/Getty Images

Endpapers: Cross-section through the 3.9-litre V12 engine developed by Cosworth for the Gordon Murray Automotive T.50 supercar – the most engaging, characterful and driver-focused V12 engine ever produced.

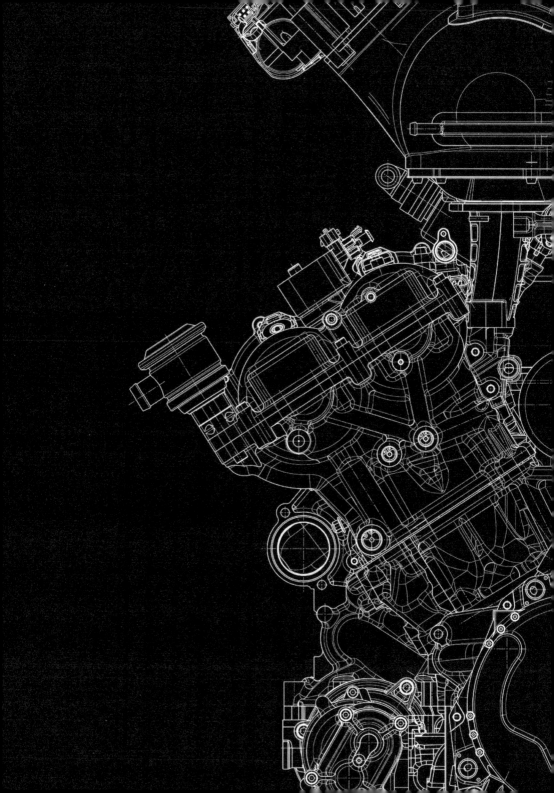